Micromechanical Investigation of Soil Deformation: Incremental Response and Granular Ratcheting

Von der Fakultät Mathematik und Physik der Universität Stuttgart
zur Erlangung der Würde eines Doktors der
Naturwissenschaften (Dr. rer. nat.) genehmigte Abhandlung

vorgelegt von

Fernando Alonso-Marroquín

aus Bogotá, Kolumbien

Hauptberichter: Prof. Dr. H. J. Herrmann
Mitberichter: Prof. Dr. P. Vermeer

Tag der mündlichen Prüfung: Februar 23, 2004

Institut für Computeranwendungen 1 der Universität Stuttgart

2004

Bibliographic information published by Die Deutsche Bibliothek

Die Deutsche Bibliothek lists this publication in the Deutsche Nationalbibliografie;
detailed bibliographic data is available in the Internet at http://dnb.ddb.de.

ISBN 3-8325-0560-1

Logos Verlag Berlin
Comeniushof, Gubener Str. 47,
10243 Berlin
Tel.: +49 030 42 85 10 90
Fax: +49 030 42 85 10 92
INTERNET: http://www.logos-verlag.de

"The main shortcoming in the field of constitutive modeling is that each researcher (or group or researchers) is developing his own constitutive model. This model is in most cases very intricate and, thus non-relocative. i.e. another researcher is unable to work with it. I can report from my own experience that it took me several months of hard work until I realized that I was unable to obtain anything with a constitutive model proposed by a colleague. How can relocativity be improved?"

D. Kolymbas 2000
The Misery of Constitutive Modeling

Publications related to this thesis

- F. Alonso-Marroquín, and H. J. Herrmann. Ratcheting of granular materials. Phys. Rev. Lett. 92, 054301 (2004).

- F. Alonso-Marroquín, and H. J. Herrmann. Calculation of the incremental stress-strain relation of a polygonal packing. Physical Review E, 66, 021301 (2002).

- F. Alonso-Marroquín, S. Luding and H. J. Herrmann. The anisotropy of granular materials. Submitted to Phys. Rev. E (2004). *cond-mat/0403064*.

- F. Alonso-Marroquín, and H. J. Herrmann. Investigation of the incremental response of soils using a discrete element model. Submitted to J. Math. Eng. (2004). *cond-mat/0403065*.

- A. Peña, A. Lizcano, F. Alonso-Marroquín, and H. J. Herrmann. Numerical simulations of biaxial test using non-spherical particles. in preparation, 2004.

- F. Alonso-Marroquín, S. McNamara, and H.J. Herrmann. Micromechanische Untersuchung des granulares Ratchetings. *DFG Antrag*, (2003).

- F. Alonso-Marroquín, R. García-Rojo, and H. J. Herrmann. Micromechanical investigation of granular ratcheting. Proceedings of International Conference on Cyclic Behaviour of Soils and Liquefaction Phenomena. Bochum-Germany (2004).

- F. Alonso-Marroquín, H. J. Herrmann, and S. Luding. Analysis of the elasto-plastic response of a polygonal packing. Proceedings of ASME International Mechanical Engineering Congress and Exposition. IMECE2002-32498 (2002).

Contents

Chapter 1

Zusammenfassung

Das mechanische Verhalten von Böden wurde mit Hilfe von Stoffgesetzen untersucht [1]. Stoffgesetze sind empirische Beziehungen, die auf Laborversuchen mit Bodenproben beruhen. Seit einigen Jahren ist es möglich, Böden auf dem Kornniveau zu untersuchen, um die mikromechanischen Aspekte von Bodendeformationen zu verstehen [2].

Ziel dieser Arbeit ist es, beide Forschungsansätze, welche die plastischen Bodendeformationen untersuchen, zusammenzubringen. Um die Auswirkung der Fabrik-Variablen [3–6], der Kettenkräfte [7, 8] und Reibungskräfte [9] auf unterschiedliche Aspekte der Bodenplastizität zu erforschen, wurde ein einfaches molekulardynamisches Modell verwendet. Herausragende Aspekte sind: Dilatanz [10], Versagen [11], Scherbands [12, 13] und Ratcheting [14].

Die Auseinandersetzung dreht sich um zwei zentrale Fragen der Bodenmechanik. Erstens: Ist die inkrementelle, nicht-lineare Theorie geeignet, um die mechanische Antwort der Böden zu beschreiben? [1, 15] Zweitens: Existiert ein ausschließlich elastischer Bereich, bei der Deformation nicht-kohäsiver Böden? [16–18].

Um die molekulardynamische Methode zu entwickeln, wurden Polygone benutzt, welche durch Kontaktkräfte wechselwirken [19, 20]. Diese Kräfte sind: Elastizität, Viskosität und Reibung [21]. Bei den biaxialen Simulationen wurden die Randbedingungen so gewählt, dass diese möglichst genau der Hüllmembran und den elastischen Wänden entsprechen, wie sie in den Versuchen verwendet werden [12].

Zur Erzeugung von polygonalen Ensembles wurde die Voronoi-Gebietzerlegungsmethode verwendet [22]. Mit dieser Methode können zahlreiche unterschiedliche Körner erzeugt werden, wie sie in wirklich existierenden Böden vorkommen. Um Beispiele mit unterschiedlichen Dichten zu generieren, begannen wir unsere Simulationsreihe mit sehr lockeren Proben, die nach und nach mit Gravitationskräften verdichtet wurden. Anschließend wurden die Proben durch zyklische Scherung solange weiterverdichtet, bis sie die gewünschte Dichte erreichten.

Biaxialversuche ergaben, dass die Dehnungslokalisierung die häufigste Art des Versagens ist. Die Hauptmerkmale des Scherbandes stimmen mit der Coulomb-Lösung überein. Das Auftreten der Dilatanz und die fortschreitende Lokalisierung der plastischen Deformation vor dem Versagen kann nicht mit solchen einfachen Methoden beschrieben werden. Bei Simulationen mit unterschiedlichen Anfangsdichten kann beobachtet werden, dass die Festigkeit und die Dichte einen konstanten kritischen Zustand annehmen. Dieser Zustand ist durch das Erreichen konstanter Schubspannungen und konstanter Dichten gekennzeichnet [23].

Ein *Repräsentatives Volumenelement* REV wurde verwendet, um die numerischen Simulationen mit den Stoffgesetzen zu vergleichen. Um die starken Schwankungen von Spannung und Deformation abzuschwächen, wurde der Durchschnitt über dieses Volumen gebildet [6]. Jedes Korn wurde als ein Teil des Kontinuums betrachtet. Unter der Annahme, dass die Spannung und Dehnung des Korns in einer kleinen Kontaktregion konzentriert sind, erhalten wir Ausdrücke für die durchschnittliche Spannung und Dehnung in Abhängigkeit von den Kontaktkräften und den individuellen Verschiebungen und Rotationen der Körner.

Die inkrementelle Spannungs-Dehnungs-Beziehung der granularen Packung wird gelöst, indem zunächst die Spannungen inkrementell mit gleicher Amplitude, aber unterschiedlichen Richtungen erhöht und dann die inkrementellen Dehnungen gemessen werden. Alle Spannungsinkremente werden auf identische polygonale Packungen angewendet. Wie die elastoplastischen Theorien vorhersagen, weisen die resultierenden Antworten zwei Tensorielle Zonen auf [15]. Wir stellen auch fest, dass das Superpositionsprinzip erfüllt ist, was im Einklang mit der Existenz dieser Tensoriellen Zonen steht. Diese Ergebnisse zeigen, dass für die Beschreibung der inkrementellen Antwort dieses Modells die Elasto-Plastizität passender ist als die inkrementellen nichtlinearen Modelle.

Die Grundelemente der elastoplastischen Theorie werden erzeugt, indem beide, die elastische und die plastische inkrementelle Antwort, berechnet werden [21]. Trotz der Einfachheit unserer Modelle können die grundlegenden Eigenschaften wirklicher Böden reproduziert werden, wie z.b. die Spannungs-Dilatanz- Beziehung [10], die nicht-assoziierte Fließregel der Plastizität [11] sowie die Existenz von Instabilitäten im Verfestigungsbereich [24].

Was die Verbindung des makromechanischen Verhaltens mit der mikromechanischen Umordnung anbelangt, können zwei wichtige Ergebnisse ermittelt werden: Erstens die Abhängigkeit der elastischen Steifigkeit von der Anisotropie der Korngerüstes [21]. Zweitens die Korrelation zwischen der plastischen Verformung und den Fabrik-Koeffizienten der gleitenden Kontakte [25]. Aus den Ergebnissen ist ersichtlich, dass die Bestimmung der Entwicklungsgleichung des Fabrik-Koeffizienten eine mikromechanische Basis der elastoplastischen Theorie darstellt.

Die Notwendigkeit einer neuen theoretischen Basis für die Bodenplastizität ergibt sich aus der Tatsache, dass einer der wichtigsten Bestandteile dieser Theorie, der elastische Bereich, mit den experimentellen Ergebnissen nicht übereinstimmt [16]. Unsere Absicht war es, diesen elastischen Bereich zu erforschen, indem zuerst die Proben belastet und dann wieder entlastet werden, um sie dann in verschiedenen Richtungen im Spannungsraum erneut zu belasten [26]. In jeder Richtung fanden wir kontinuierliche Übergänge von elastischem zu elastoplastischem Verhalten, so dass wir kein rein elastisches Regime identifizieren konnten. Auf mikromechanischer Ebene ist es ersichtlich, dass dieser Effekt von der Tatsache herrührt, dass jede Ladung gleitende Kontakte beinhaltet.

Dadurch, dass das elastische Regime vernachlässigbar klein wurde, stellte sich das hysteretische Verhalten als ein überraschender Aspekt der Untersuchungen heraus [27]. Bei quasi-statischer, zyklischer Belastung folgt auf das hysteretische Verhalten eine schrittweise Akkumulation der plastischen Verformung mit der Zyklenzahl. Eine numerische Simulation von Schubspannungszyklen mit Amplituden zwischen $0.001p$ und $0.6p$ (p ist der Seitendruck) zeigt einen asymptotischen Verformungsverlauf, wobei eine konstante Verformungszunahme pro Zyklus auftritt. Das durch die Zustandstandsvariablen Spannung und Porenzahl beschriebene System erreicht dabei niemals einen kritischen Zustand. Dieser unerwartete Effekt, der granulares Ratcheting genannt wird, kann nicht mit den gängigen elasto-plastischen Konzepten interpretiert werden. Die elastoplastische Theorie besagt, dass für Ladezyklen unterhalb eines bestimmten Wertes, der als Shakedown Limit bezeichnet wird, die Akkumulation der plastischen Deformation nach einer bestimmten Anzahl von Zyklen zum Erliegen kommt [28].

Das granulares Ratcheting wird aus mikromechanischer Sicht durch die Beobachtung des Verlaufs der Mikrokontaktkräfte bei quasistatischer Belastung des granularen Materials untersucht. Unsere Berechnungen zeigen, dass jede Deformation, die auf den Rand wirkt, sich heterogen in der Probe verteilt. Wenn die Probe isotrop komprimiert wird, erreichen einige Kontaktkräfte das Coulombsche Reibungskriterium $|f_t| = \mu f_n$. Dies führt zu irreversiblen Verformungen im Korngerüst.

Zwischen der Steifigkeit der Probe und der Anzahl der gleitenden Kontakte wird eine hohe Korrelation bei zyklischer Belastung beobachtet. Tatsache ist, dass während des Überganges von Belastung zu Entlastung eine abrupte Abnahme der Zahl der gleitenden Kontakte auftritt. Als Ergebnis lässt sich ein Sprung in der Steifigkeit beobachten. Hierbei ist die Steifigkeit unter Entlastung größer als unter Belastung. Andererseits lässt sich während der Belastungszyklen eine kontinuierliche Abnahme der Steifigkeit beobachten. Dies zeigt die Abhängigkeit der Steifigkeitsabnahme von der wachsende Anzahl der gleitenden Kontakte.

Bei hinreichend kleiner Amplitude der zyklischen Belastung können die bleibenden Verformungen der Probe durch das wiederholt erreichte Reibungskriterium der Kontakte beschrieben werden. Über lange Zeiträume verhält sich eine kleine Menge der Kontakte wie Rätschen. Diese erreichen periodisch das Reibungskriterium und verursachen irreversible Verformungen im Material. Sie rutschen in jeder Belastungsphase, und verhalten sich elastisch in den Entlastungsphasen.

Dieses Ergebnis legt nahe, dass die plastische Verformung des Bodens mit Hilfe geeigneter Statistiken über die Reibungskontakte beschrieben werden kann. Diese Statistik kann formal in das Stoffgesetz eingebunden werden, indem geeignete Strukturtensoren als Zustandsgrößen des Korngerüsts eingeführt werden. Diese Strukturtensoren würden eine mikromechanische Interpretation der Stoffgesetzes liefern.

Chapter 2

Introduction

The 1960s was significant for the development of soil mechanics and, in particular, the constitutive models for soils. Prior to this decade, soil mechanics was confined to linear elastic theory [29] and the Mohr-Coulomb failure criterion [30]. A radical change of the perspectives of soil plasticity occurred after the pioneering work of Roscoe and his coworkers in Cambridge, which led to the basic principles of the Critical State Theory [31]. The prototype of this theory was the so-called Cam-Clay model [32]. With five material constants, this model was the first nonlinear representation describing several aspects of deformation and failure of soils.

In an attempt to cover further aspects of the cyclic soil behavior, subsequent developments have given rise to a great number of constitutive models [1]. Unfortunately these models give only satisfactory results in the small range of experiments where they were developed. Other models attempting to represent a wider range of phenomena had to incorporate a large number of parameters. These parameters not only lack physical meaning, but are also very difficult to calibrate with the experimental data.

This tendency to increase the number of constants in the models has been pointed out by Scott in the workshop *Constitutive Equations for Granular Noncohesive Soils* in 1988 [33]. In this meeting models with up to 40 calibration parameters were presented. By performing a survey on the constitutive relations presented in previous international workshops, he reported that the number of constants was growing at about 12% per year. Extrapolating this observation, he estimated that models developed in the year 2000 would have 184 constants!

In contradiction to Scott's predictions, no model has been reported with this many material parameters until now. However, the large number of concepts that have been introduced has driven a proliferation of constitutive models [34]. The strong controversy concerning the validity of a large number of models and the lack of experimental meaning of the material parameters has led the practitioners to lose confidence in constitutive modeling. This has resulted in a gap between research and practice in geotechnical engineering [35].

In geotechnical applications, it is desirable that the parameters of a constitutive relation depend directly on the properties of the grains. In the simple case of dry soils, granulometric properties can involve grain shape and angularity, distribution of grain sizes, friction coefficient and stiffness of the grains [36]. Unfortunately, the existing models do not consider these granulometric properties, but employ unfamiliar abstract parameters instead.

An alternative for the investigation of soils at the grain scale is the discrete element modeling (DEM) [37]. Examples of this approach are the contact mechanics method (CM) [9, 38] and the molecular dynamics (MD) [39]. These discrete approaches take into account details like particle shape, size distribution, friction and cohesion between the grains. The interaction between the particles is modeled by the introduction of suitable contact forces. These forces are given in terms of a reduced number of parameters. The MD method introduces the normal and tangential stiffnesses, and the friction coefficients as the material constants of the grains. In CD the particles are supposed to be infinitely rigid, and the interactions between the grains are described by a Coulomb friction law with a single friction coefficient.

Disks or spheres are used in order to capture the granularity of the materials [5, 8, 37, 40]. The simplicity of their geometry allows one to reduce the computer time of calculations. However, they do not take into account the diversity of grain shapes in realistic materials. A more detailed description with three-dimensional non-spherical particles has been presented [41], but the applicability of these models is still limited by the computational time of the simulations.

The comparison of the simulation with the constitutive theories requires a *homogenization* technique. This is a formalism that allows us to derive macromechanical quantities from micromechanical variables. Different homogenization techniques have been used to derive the stress [17, 42, 43] and the strain tensor [4, 44–46]. Although the different homogenization approaches converge to the same micromechanical expression for the stress tensor, the micromechanical definition of the strain tensor is still under discussion.

From the derivation of the stress-strain relation one can bridge the gap between the discrete and continuum approaches. The incremental theory provides a simple method to obtain the incremental stress-strain relation directly from DEM simulations without recourse to any particular constitutive model [24]. This method has been used to calculate the incremental response of disks [40] and spheres [47, 48]. Some recent results seem to contradict many well-established concepts of the elasto-plastic theory [48, 49]. However, it should be addressed that the behavior of spherical packings is qualitatively different from realistic soil samples. In particular, it has been shown that the friction angle of a packing of spheres is much lower than the experimental values for sand [49]. This is given by the fact that a sphere can rotate much more easily inside a packing rather than an arbitrarily shaped grain. It is, therefore, of obvious interest to study the incremental response of non-spherical particles.

In this work we perform MD simulations using a simplified model, where the particles are represented by randomly shaped polygons. This model will be applied to perform a micromechanical investigation of plastic deformation of soils. In reality, the plastic deformation of granular materials is produced by grain rearrangements and grain crushing. In our model we assume that the grains cannot break, and we take into account only of the role of the sliding contacts in the plastic deformation of the granular assembly.

This work is organized as follows: In Chapter 3 we introduce the basic ingredients of the model. Chapter 4 considers the biaxial test, which is discussed in the frame of the Mohr-Coulomb criterion and the Critical State Theory. In Chapter 5 we introduce an homogenization procedure, which will be used to calculate the incremental relation of the models. The elasto-plastic and the hypoplastic features of the incremental response are discussed in this chapter. In Chapter 6 the stiffness tensor and the flow rule are calculated from the resulting incremental response. The constitutive response is obtained in terms of some internal variables, which take into account the anisotropy induced by the loading on the contact network. We also explore two basic concepts of elasto-plasticity: the Hill condition of stability and the question of the existence of an elastic regime.

In Chapter 7 we investigate the response of dense polygonal samples when they are subjected to load-unload stress cycles. The accumulation of permanent deformation and the compaction of the sample are studied as a function of the number of cycles, taking different loading amplitudes. We report on the existence of ratcheting regimes for extremely small loading amplitudes. This ratcheting is studied at the grain level, following the evolution of the contact forces, and the kinematics of the individual grains. We also investigate the correlation between the hysteretic behavior of the stiffness and the evolution of the sliding contacts.

Chapter 3

The Model

In this chapter, an extension of the two-dimensional discrete element methods that have been used to model granular materials via polygonal particles [19, 50] are presented. The model captures many aspects of realistic granular materials, such as the elasticity, friction, damping forces and the possibility of slippage. Boundary conditions are introduced to model surrounding flexible membranes and rigid walls. Using a simplified method of random generation of polygons, we are able to capture an important aspect of granular soils that is the diversity of shapes and sizes of the grains.

Of course, there are some limitations in the use of such a two-dimensional model to study physical phenomena that are three-dimensional in nature. These limitations have to be kept in mind in the interpretation of the results and its comparison with the experimental data. In order to give a three-dimensional picture of this model, one can consider the polygons as a collection of prismatic bodies with randomly-shaped polygonal basis. Alternatively, one could consider the polygons as three-dimensional grains whose centers of mass all move in the same plane. It is the author's opinion that it is more sensible to consider this model as an idealized granular material that can be used to check the constitutive models.

The details of the particle generation, the contact forces, the boundary conditions and the molecular dynamics simulations are presented in the following sections.

3.1   Generation of polygons

The polygons representing the particles in this model are generated by using the method of Voronoi tessellation [50]. First, a regular square lattice of side ℓ is created. Next, we set a random point in a square of side length a inside the cells of the rectangular grid. Then, each polygon is constructed, assigning to each point that part of the plane that is nearer to it than to any other point. The details of the construction of the Voronoi cells can be found in the literature [22, 51]. The tessellations can also be implemented by using standard programs, such as Matlab. The details of the construction are skipped, and the most salient geometrical aspects of these Voronoi constructions are presented.

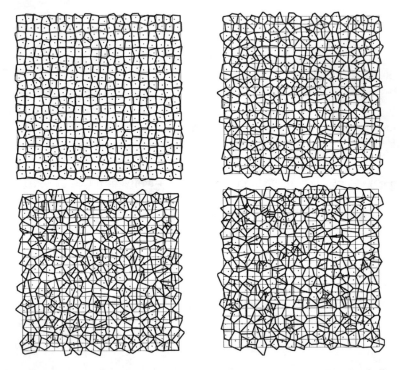

Figure 3.1: Voronoi construction used to generate the convex polygons. The dots indicate the point used to the tessellation. Periodic boundary conditions were used. Four different values of a are chosen: 0.5ℓ, $\ell, 2\ell$ and 20ℓ.

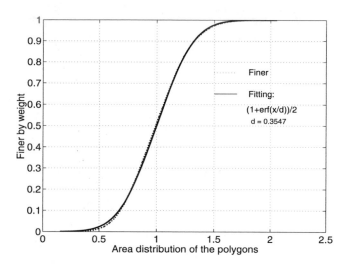

Figure 3.2: Cumulative distribution of polygon areas. The solid line shows the fit of the data using an error function. The distribution is calculated for 1.8×10^4 polygons generated with $a = \ell$.

Fig. 3.1 shows random tessellations for different values of a. The tessellation with $a \leq \ell$ corresponds to the so-called *vectorizable random lattices* [22]. They are Voronoi constructions with low disorder, a narrow distribution of areas and a certain anisotropy when $a < \ell$ [52]. This anisotropy is reflected in the fact that the orientational distribution of the edges is not uniform. The computational advantage of the constructions with $a \leq \ell$ is that the number of potential neighbors of each polygon is bounded to 20 [22]. This property allows one to fix the neighbor list during the simulation, which reduces the time required to calculate the interactions between the polygons [19].

The tessellations with $a > \ell$ lead to isotropic Voronoi tessellations with a wide, asymmetric distributions of areas of the polygons. In particular, the limit $a \gg \ell$ corresponds to the so-called Poisson tessellations [22, 51]. In the case where $a = \ell$ the orientational distribution of edges is isotropic, and the diversity of areas of polygons is symmetric around ℓ^2, as shown the Fig. 3.2. These two properties are observed in natural river sand [12, 53]. The probabilistic distribution of areas follows approximately a Gaussian

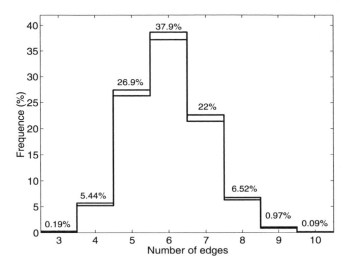

Figure 3.3: Distribution of number of edges. Five different random tessellations of 60×60 cells with $a = \ell$ were used in the calculations. The bars show the standard deviation of the data.

distribution with a variance of $0.36\ell^2$. Voronoi tessellations with $a = \ell$ will be used in this work.

Using the Euler theorem, it has been shown analytically that the mean number of edges of any random tessellation must be 6 [51]. Further statistical measures have not been analytically derived and they have to be estimated numerically [22]. The distribution of number of edges of the polygons has been numerically calculated here in the case $a = \ell$ using 5 different random tessellations of 60×60 cells. We found that the number of edges is distributed between 4 and 8 for 98.7% of the polygons, as shown Fig. 3.3.

Note that volume fraction of this Voronoi construction is one. This perfect packing is an unrealistic condition of granular materials. In order to have packing fractions lower than one we use a procedure which is explained in Sec. 4.4.

3.2   Contact forces

In order to calculate the forces, we assume that all the polygons have the same thickness L. The force between two polygons is written as $\mathbf{F} = \mathbf{f}L$ and the mass of the polygons is $M = mL$. In reality, when two elastic bodies come into contact, they have a slight deformation in the contact region. In the calculation of the contact force we suppose that the polygons are rigid, but we allow them to overlap. Then, we calculate the force from this virtual overlap.

The first step for the calculation of the contact force is the definition of the line representing the flattened contact surface between the two bodies in contact. This is defined from the contact points resulting from the intersection of the edges of the overlapping polygons. In most cases, we have two contact points, as shown in the left of Fig. 3.4. In such a case, the contact line is defined by the vector $\mathbf{C} = \overrightarrow{C_1C_2}$ connecting these two intersection points. In some pathological cases, the intersection of the polygons leads to four or six contact points. In these cases, we define the contact line by the vector $\mathbf{C} = \overrightarrow{C_1C_2} + \overrightarrow{C_3C_4}$ or $\mathbf{C} = \overrightarrow{C_1C_2} + \overrightarrow{C_3C_4} + \overrightarrow{C_5C_6}$, respectively. This choice guarantees a continuous change of the contact line, and therefore of the contact forces, during the evolution of the contact.

The contact force is separated as $\mathbf{f}^c = \mathbf{f}^e + \mathbf{f}^v$, where $\mathbf{f}^e$ and $\mathbf{f}^v$ are the elastic and viscous contribution. The elastic part of the contact force is decomposed as $\mathbf{f^e} = f_n^e \hat{n}^c + f_t^e \hat{t}^c$. The calculation of these components is explained below. The unit tangential vector is defined as $\hat{t}^c = \mathbf{C}/|\mathbf{C}|$, and the normal unit vector $\hat{n}^c$ is taken perpendicular to $\mathbf{C}$. The point of application of the contact force is taken as the center of mass of the overlapping polygon.

3.2.1   Normal elastic force

As opposed to the Hertz theory for round contacts, there is no exact way to calculate the normal force between interacting polygons. An alternative method has been proposed in order to model this force [19]. In this method, the normal elastic force is calculated as $f_n^e = -k_n A/L_c$ where k_n is the

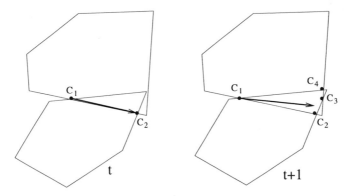

Figure 3.4: Contact points C_i before (left) and after the formation of a pathological contact (right). The vector denotes the contact line. t represents the time step.

normal stiffness, A is the overlapping area and L_c is a characteristic length of the polygon pair. Our choice of L_c is $1/2(1/R_i + 1/R_j)$ where R_i and R_j are the radii of the circles of the same area as the polygons. This normalization is necessary to be consistent in the units of force [50].

3.2.2 Frictional forces

In order to model the quasistatic friction force, we calculate the elastic tangential force using an extension of the method proposed by Cundall and Strack [37]. An elastic force $f_t^e = -k_t \Delta x_t$ proportional to the elastic displacement is included at each contact. k_t is the tangential stiffness. The elastic displacement Δx_t is calculated as the time integral of the tangential velocity of the contact during the time when the elastic condition $|f_t^e| < \mu f_n^e$ is satisfied. The sliding condition is imposed, keeping this force constant when $|f_t^e| = \mu f_n^e$. The straightforward calculation of this elastic displacement is given by the time integral starting at the beginning of the contact:

$$\Delta x_t^e = \int_0^t v_t^c(t')\Theta(\mu f_n^e - |f_t^e|)dt', \qquad (3.1)$$

where Θ is the Heaviside step function and v_t^c denotes the tangential component of the relative velocity $\mathbf{v}^c$ at the contact:

$$\mathbf{v}^c = \mathbf{v}_i - \mathbf{v}_j - \boldsymbol{\omega}_i \times \boldsymbol{\ell}_i + \boldsymbol{\omega}_j \times \boldsymbol{\ell}_j. \tag{3.2}$$

Here $\mathbf{v}_i$ is the linear velocity and $\boldsymbol{\omega}_i$ is the angular velocity of the particles in contact. The branch vector $\boldsymbol{\ell}_i$ connects the center of mass of particle i with the point of application of the contact force.

3.2.3 Damping forces

Damping forces are included in order to allow for rapid relaxation during the preparation of the sample, and to reduce the acoustic waves produced during the loading. These forces are calculated as

$$\mathbf{f}_v^c = -m(\gamma_n v_n^c \hat{n}^c + \gamma_t v_t^c \hat{t}^c), \tag{3.3}$$

being $m = (1/m_i + 1/m_j)^{-1}$ the effective mass of the polygons in contact, $\hat{n}^c$ and $\hat{t}^c$ are the normal and tangential unit vectors defined before, and γ_n and γ_t are the coefficients of viscosity. These forces introduce time dependent effects during the cyclic loading. However, we will show that these effects can be arbitrarily reduced by increasing the time of loading, as corresponds to the quasistatic approximation.

In order to solve the equations of motion, it is necessary to specify the forces acting on the particles on the boundary. Two different boundary conditions are used in the calculations. The *floppy boundary* method allows one to perform a stress-controlled test on the sample without imposing any restriction on the deformation of the assembly. Elastic walls can also be used to control the deformation of the polygonal assembly. These two boundary conditions are presented in the following sections.

3.3   Floppy boundary

The method of floppy boundary is introduced to model the typical biaxial test used to investigate the strain localization [53]. In this test, a prismatic granular sample, surrounded by a latex membrane, is placed between two fixed walls to create plane strain condition. Then the sample is subjected to axial loading, superimposed by a confining pressure applied on the membrane.

We are going to discuss how the latex membrane can be modeled. One way would be to apply a perpendicular force on each edge of the polygons belonging to the external contour of the sample. Actually, this does not work because the force will act on all the fjords of the boundary. This produces an uncontrollable growth of cracks that with time, end up destroying the sample. With a latex membrane this cannot happen because the bending stiffness of the membrane does not allow the pressure to penetrate in all the fjords of the sample. To model such a membrane, we will introduce a criterion which restricts the boundary points that are subjected to the external stress.

The algorithm to identify the boundary is rather simple. The lowest vertex p from all the polygons of the sample is chosen as the first point of the boundary list b_1. In Fig. 3.5 P is the polygon that contains p, and $q \in P \cap Q$ is the first intersection point between the polygons P and Q in counterclockwise orientation with respect to p. Starting from p, the vertices of P in counterclockwise orientation are included in the boundary list until q is reached. Next, q is included in the boundary list. Then, the vertices of Q between q and the next intersection point $r \in Q \cap R$ in the counterclockwise orientation are included in the list. The same procedure is applied until one surrounds the sample and reaches the lowest vertex p again. This is a very fast algorithm, because it only makes use of the contact points between the polygons, which are previously calculated to obtain the contact force in each time step.

Let's define $\{b_i\}$ the set of points of the boundary and $\{m_i\}$ the set of boundary points that are in contact with the membrane. They are selected using a recursive algorithm. It is initialized with the vertices of the smallest convex polygon that encloses the boundary (see Fig. 3.6). The lowest point

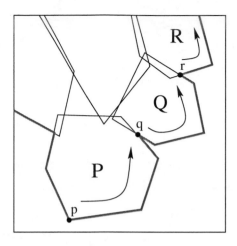

Figure 3.5: Algorithm used to find the boundary.

of the boundary is selected as the first vertex of the polygon $m_1 = b_1$. The second one m_2 is the boundary point b_i that minimizes the angle $\angle(\overrightarrow{b_1 b_i})$ with respect to the horizontal. The third one m_3 is the boundary point b_i such that the angle $\angle(\overrightarrow{m_2 b_i}, \overrightarrow{m_1 m_2})$ is minimal. The algorithm is recursively applied until the lowest vertex m_1 is reached again.

The points of the boundary are iteratively included in the list $\{m_i\}$ using the bending criterion proposed by Åstrøm [54]. For each pair of consecutive vertices of the membrane $m_i = b_i$ and $m_{i+1} = b_j$ we choose that point from the subset $\{b_k\}_{i \leq k \leq j}$ which maximizes the bending angle $\theta_b = \angle(\overrightarrow{b_k b_i}, \overrightarrow{b_k b_j})$. This point is included in the list whenever $\theta_b \geq \theta_{th}$. Here θ_{th} is a threshold angle for bending. This algorithm is repeatedly applied until there are no more points satisfying the bending condition. The final result gives a set of segments $\{\overrightarrow{m_i m_{i+1}}\}$ lying on the boundary of the sample as shown in Fig. 3.6.

In order to apply stress at the boundary, the segments of the membrane are divided into two groups: A-type segments are those that coincide with an edge of a boundary polygon; B-type segments connect the vertices of two different boundary polygons. On each segment of the membrane $\mathbf{T} = \Delta x_1 \hat{x}_1 + \Delta x_2 \hat{x}_2$, we apply a force

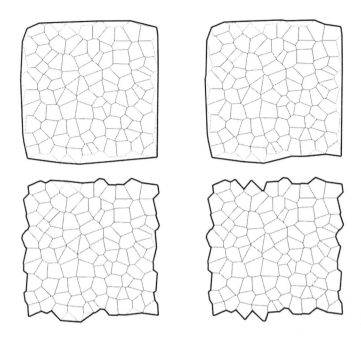

Figure 3.6: Floppy boundary obtained with threshold bending angle $\theta_{th} = \pi, 3\pi/4, \pi/2$ and $\pi/4$, the first one corresponds to the minimum convex polygon that encloses the sample.

$$\mathbf{f}_s^m = -\sigma_1 \Delta x_2 \hat{x}_1 + \sigma_2 \Delta x_1 \hat{x}_2 \tag{3.4}$$

Here $\hat{x}_1$ and $\hat{x}_2$ are the unit vectors of the Cartesian coordinate system. σ_1 and σ_2 are the components of the stress we want to apply on the sample. This force is transmitted to the polygons in contact with it. If the segment is A-type, this force is applied at its midpoint; if the segment is B-type, half of the force is applied at each one of the vertices connected by this segment. An additional damping force $\mathbf{f}_v^m = -\gamma_b m_i \mathbf{v}^b$ is included to reduce the acoustic waves produced during loading. Here γ_b is the coefficient of viscosity of the floppy boundary, and m_i is the mass of the polygon and $\mathbf{v}^b$ the velocity of the polygon.

This boundary condition has been used in Chapter. 4 to simulate biaxial tests. We have observed shear bands whose orientation seems to be sensitive to the threshold bending angle θ_{th}. However, some problems have been detected in the use of such bending conditions. For small values of θ_{th} the floppy boundary penetrates too much in the fjords, producing some instabilities in the boundary polygons. This instability is reflected in large displacements on boundary polygons for small loadings, eventually leading to their detachment.

Moreover, for values of θ_{th} close to π we have detected numerical problems. When the sample is kept at constant isotropic pressure, the assembly cannot reach an equilibrium configuration. We have observed that in these cases the floppy boundary flips periodically to different configurations, giving rise to spurious oscillations in the assembly. A reason for this numerical problem could be the fact that this method leads to boundary forces which do not change continuously with time. In these cases, the numerical method used to solve the equations of motion cannot guarantee stability and convergence of numerical solutions.

3.4   Walls as boundaries

Usually, the granular assemblies are compacted and loaded within a set of confining walls. These walls act as boundary conditions, and can be moved by specifying their velocity or the force applied on them. The response of the walls can be used to calculate the global stress and strain of the assembly.

The interaction of the polygons with the walls is modeled here by using a simple visco-elastic force. First, we allow the polygons to penetrate the walls. Then, for each vertex of the polygon α inside of the walls we include a force

$$\mathbf{f}^b = -k_n \delta \mathbf{n} - \gamma_b m_\alpha \mathbf{v}^b, \qquad (3.5)$$

where δ is the penetration length of the vertex, $\mathbf{n}$ is the unit normal vector to the wall, and $\mathbf{v}^b$ is the relative velocity of the vertex with respect to the wall.

3.5 Molecular dynamics simulation

The evolution of the position $\mathbf{x}_i$ and the orientation φ_i of the polygon i is governed by the equations of motion:

$$m_i\ddot{\mathbf{x}}_i = \sum_c \mathbf{f_i^c} + \sum_b \mathbf{f}_i^b + \sum_m \mathbf{f}_i^m,$$

$$I_i\ddot{\varphi}_i = \sum_c \boldsymbol{\ell}_i^c \times \mathbf{f_i^c} + \sum_b \boldsymbol{\ell}_i^b \times \mathbf{f}_i^b + \sum_m \boldsymbol{\ell}_i^m \times \mathbf{f}_i^m. \qquad (3.6)$$

Here m_i and I_i are the mass and moment of inertia of the polygon. The first sum goes over all those particles in contact with this polygon; the second one over all the vertices of the polygon in contact with the walls, and the third one over all the edges in contact with the floppy boundary. $\mathbf{f}^m$ and $\mathbf{f}^b$ are the forces applied on the polygons in contact with the floppy boundary and the walls, respectively, which where defined in Sec. 3.3 and Sec. 3.4. The interparticle contact forces $\mathbf{f^c}$ are given by

$$\mathbf{f^c} = -(k_n A/L_c + \gamma_n m v_n^c)\mathbf{n}^c - (\Delta x_t^c + \gamma_t m v_t^c)\mathbf{t}^c,$$

$$(3.7)$$

where A is the overlapping area of the interacting polygons and L_c the characteristic length of the contacts, both defined in Subsect. 3.2.1 Δx_t^e denotes the elastic part of tangential displacement of the contact, which were defined in Sec. 3.2.2. σ_i^b is the stress applied on the boundary segment with normal vector $\mathbf{N}_b$. The effective mass m of the polygons, the coefficient of viscosity γ_n and γ_t, and the relative velocity at the contact $\mathbf{v}^c$ are defined in Sec. 3.2.3.

We use a fifth-order Gear predictor-corrector method for solving the equation of motion [39]. This algorithm consists of three steps. The first step predicts position and velocity 0f the particles by means of a Taylor expansion. The second step calculates the forces as a function of the predicted

Symbol	Default value	Parameter
kn	$160MPa$	normal contact stiffness
kt	$52.8MPa$	tangential contact stiffness
μ	0.25	friction coefficient
γ_n	$4 \times 10^3 s^{-1}$	normal coefficient of viscosity
γ_t	$8 \times 10^2 s^{-1}$	tangential coefficient of viscosity
γ_b	$4 \times 10^1 s^{-1}$	coefficient of viscosity of the walls
t_0	$0.1s$	time of load
dt	$2.5 \times 10^{-6} s$	time step for the molecular-dynamics
ρ	$1gr/cm^3$	density of the grains
ℓ	$1.0cm$	size of the cells of the Voronoi generation
p_0	$160KPa$	confining pressure
θ_{th}	$\pi/4$	bending angle of the floppy boundary

Table 3.1: Parameters of the Molecular dynamics simulations.

positions and velocities. The third step corrects the positions and velocities in order to optimize the stability of the algorithm. This method is much more efficient than the simple Euler approach or the Runge-Kutta method, especially for problems where very high accuracy is a requirement.

3.6 Determination of the parameters

The parameters of the molecular dynamics simulations were adjusted according to the following criteria: 1) guarantee the stability of the numerical solution, 2) optimize the time of the calculation, and 3) provide a reasonable agreement with the experimental data.

There are many parameters in the molecular dynamics algorithm. Before choosing them, it is convenient to make a dimensional analysis. In this way, we can keep the scale invariance of the model and reduce the parameters to a minimum of dimensionless constants.

As shown in Table 3.1, there are 2 dimensionless and 10 dimensional parameters. The latter ones can be reduced by introducing the following characteristic times of the simulations: the loading line t_0, the relaxation times $t_n = 1/\gamma_n$, $t_t = 1/\gamma_t$, $t_b = 1/\gamma_b$ and the characteristic period of oscillation $t_s = \sqrt{k_n/\rho\ell^2}$ of the normal contact.

Using the Buckingham Pi theorem [55], one can show that the strain response, or any other dimensionless variable measuring the response of the assembly during loading, depends only on the following dimensionless parameters: $\alpha_1 = t_n/t_s$, $\alpha_2 = t_t/t_s$, $\alpha_3 = t_b/t_s$, $\alpha_4 = t_0/t_s$, the ratio $\zeta = k_t/k_n$ between the stiffnesses, the friction coefficient μ and the ratio p_0/k_n between the confining pressure and the normal stiffness.

The variables α_i will be called *control parameters*. They are chosen in order to satisfy the quasistatic approximation, i.e. the response of the system does not depend on the loading time, but a change of these parameters within this limit does not affect the strain response. $\alpha_1 = 0.1$ and $\alpha_2 = 0.5$ were taken large enough to have a high dissipation, but not too large to keep the numerical stability of the method. $\alpha_3 = 0.001$ is chosen by optimizing the time of consolidation of the sample in Sec. 4.4. The ratio $\alpha_4 = t_0/t_s = 10000$ was chosen large enough in order to avoid rate-dependence in the strain response, corresponding to the quasistatic approximation. Technically, this is performed by looking for the value of α_4 such that a reduction of it by half makes a change of the stress-strain relation less than 5%.

The parameters ζ and μ can be considered as *material parameters*. They determine the constitutive response of the system, so they must be adjusted to the experimental data. In this study, we have adjusted them by comparing the simulation of biaxial tests of perfect polygonal packings to the corresponding tests with dense Hostun sand [53]. First, the initial Young modulus of the material is linearly related to the value of normal stiffness of the contact. Thus, $k_n = 160 MPa$ is chosen by fitting the initial slope of the stress-strain relation in the biaxial test. Then, the Poisson ratio depends on the ratio $\zeta = k_t/k_n$. Our choice $k_t = 52.8 MPa$ gives an initial Poisson ratio of 0.07. This is obtained from the initial slope of the curve of volumetric strains versus axial strain. The angles of friction and the dilatancy are increasing functions of the friction coefficient μ. Taking $\mu = 0.25$ yields a angles of friction of $30 - 40$ degrees and dilatancy angles of $20 - 30$ degrees. The experimental data yields angles of friction between $40 - 45$ degrees and dilatancy angles between $7 - 14$ degrees. A better adjustment would be made by including different void ratios in the simulations, but this is beyond of the scope of this work.

Chapter 4

Biaxial test

An interesting phenomenon in pressure confined granular materials is that the deformation under shearing is not homogeneous, but rather is concentrated in thin layers of intensive shearing [53]. This phenomenon has attracted the attention of many researchers due to the fact that the classical continuum theories lead to ill-posed mathematical problems [56]. Different regularization approaches have been proposed, pointing to the necessity to introduce the effect of the microstructure in the continuum relations. These new models lead to certain theoretical predictions, which deserve experimental corroboration.

In some recent studies, numerical simulations of sheared packing of disks, spheres and polygons has also shown this localization of strain [57]. Numerical experiments on simple ring shear show shear bands arising for large deformation, having a characteristic width in terms of grain diameters [58]. Shear bands have also been observed in numerical simulations of the biaxial test of disk packing [59]. Particle rotations, which eventually give place to rotating bearings, is one of the major factors controlling the dilatancy and failure of these discrete models. However, these free rotations are far from the realistic micromechanical arrangements of soils, where the nonsphericity of the contacts lead to an important contribution of the slippage at the contacts to the total deformation of the granular assembly.

In this chapter we study this strain localization using molecular dynamics simulations of a dense packings of polygons. The boundary conditions are chosen in order to mimic the experimental setup of the biaxial test [53]. The latex membrane surrounding the granular sample is modeled by using the method of floppy boundary explained in the last chapter. The axial stress is controlled by moving two horizontal walls with constant velocity. The results are used to evaluate the extent of the validity of the Mohr-Coulomb failure criterion and the Critical State Theory.

This chapter is organized as follows. In Sec. 4.1 we introduce the basic elements of the Mohr-Coulomb theory, which is used to describe the onset of plastic deformation of soils. In Sec. 4.2 we present the simulation results of the quasistatic loading with axial strain control. In Sec. 4.3 we present some micromechanical aspects of the hardening process. Finally, Sec. 4.4 concerns the effects of the initial density on the response of the polygonal packing.

4.1   Mohr-Coulomb analysis

The simplest description of the stability and failure of granular materials is given by the Mohr-Coulomb criterion [11]. The basic assumption of this theory is that the granular material behaves perfectly elastic, except in the case where the normal σ_n and deviatoric σ_t stress components on a plane satisfy the failure criterion:

$$\sigma_t = c + \sigma_n tan(\varphi), \tag{4.1}$$

This is given by two material constants: *angle of friction* φ and the *coefficient of cohesion* c. The examination of the failure limit in the biaxial test is performed by taking the principal values σ_1 and σ_2 of stress of a volume element, as shown in the left side of Fig. 4.1. Let's divide the element into two pieces, separated by a plane with inclination angle θ. The equilibrium condition of one of these pieces leads to:

$$\begin{aligned} \sigma_n &= p + q\cos(2\theta), \\ \sigma_t &= q\sin(2\theta) \end{aligned} \tag{4.2}$$

where $p = (\sigma_1 + \sigma_2)/2$ and $q = (\sigma_1 - \sigma_2)/2$ are the pressure and deviatoric stress components of the volume element. According to this equation, in a diagram of σ_n versus σ_t, the stresses applied on this plane are represented by a point in the circle with radius q centered at $(p, 0)$ with an inclination angle 2θ. The failure condition of Eq. (4.1) is represented in this diagram by a cone with angle φ and a vertex located at $(-c \cot \varphi, 0)$. This construction is shown in the right part of the Fig. 4.1. According to this, the failure is reached when the circle touches the cone, and the failure plane has an angle of orientation θ_C satisfying $2\theta_C = 180° - \beta$, where β is defined in Fig. 4.1. Since $\varphi + \beta = 90°$, the orientation of the failure plane results:

$$\theta_C = 45° + \varphi/2 \tag{4.3}$$

Using the triangle from the left part of Fig. 4.1, one can obtain the material parameters from the stress components p and q of the sample at failure:

$$\sin \varphi = \frac{q}{p + c} \tag{4.4}$$

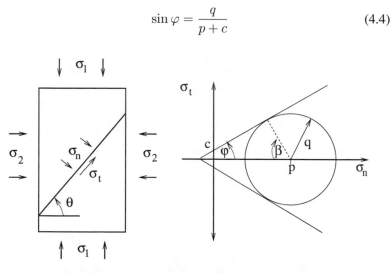

Figure 4.1: Mohr-Coulomb analysis of failure. Left: stress acting on a plane of the volume element. Right: Mohr-Coulomb circle and failure cone.

4.2   Simulation results

The analysis of failure is performed here by simulating biaxial tests on polygonal packings. First, a confining pressure is applied to the sample through the floppy boundary. Then, two horizontal walls at the top and bottom of the packing are used to apply vertical loading with constant velocity. The stress is calculated from the forces applied on the floppy boundary as $\sigma_{ij} = \frac{1}{V}\sum_b f_i^b x_j^b$, where $\mathbf{x}^b$ is the point of application of the boundary force $\mathbf{f}^b$ and V is the area enclosed by the floppy boundary [37]. From the principal values of this tensor, one can define the pressure and the deviatoric stress as $p = (\sigma_1 + \sigma_2)/2$ and $q = (\sigma_1 - \sigma_2)/2$. The axial strain is calculated as $\epsilon_1 = \Delta H/H_0$, where H is the height of the sample. The volumetric strain is given by $\epsilon_v = \Delta V/V_0$, where V is the area enclosed by the floppy boundary.

The evolution of the deviatoric stress and the volumetric strain are shown in Fig. 4.2 for different confining pressures. The strain response is characterized by a continuous decrease of the stiffness, i.e. the slope of the stress-strain curve, from the very beginning of the load process. The failure is given by the peak stress value (i.e. the maximal stress reached during the loading). The volumetric strain has a compaction regime from the beginning of the load, and dilatancy before failure. The maximal dilatancy is observed around the failure. For large loadings, the sample reaches a stationary state where the stress and the volume remain approximately constant, except for some fluctuations which remain for large deformations.

An important remark is that the Mohr-Coulomb criterion does not provide a complete description of the failure. First, the volume expansion should be an integral part of this description. Second, the relation between the pressure and the deviatoric stress at failure shows slight deviations from the Mohr-Coulomb theory. As shown in Fig. 4.3, they are not related linearly, but they approximately satisfy a power law

$$\frac{p}{p_r} = \alpha(\frac{q}{p_r})^\beta, \tag{4.5}$$

where $p^r = 1MPa$, $\alpha = 0.625$ and $\beta = 0.93$. An interesting consequence

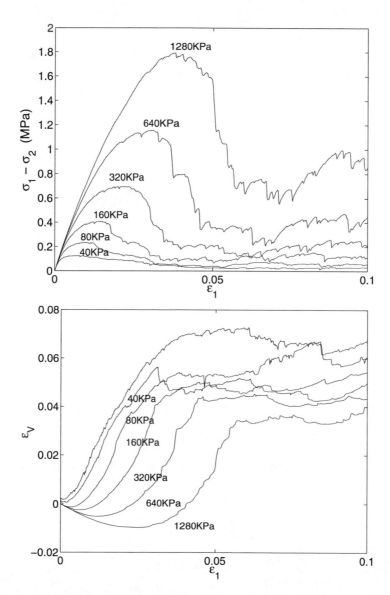

Figure 4.2: Deviatoric stress and volumetric strain versus axial strain for different confining pressures.

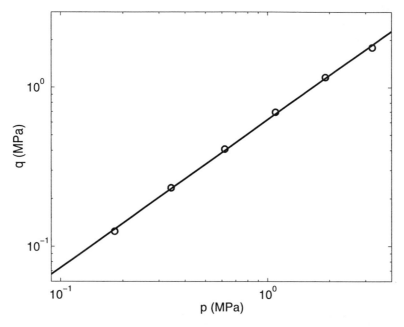

Figure 4.3: Relation between the deviatoric stress and the pressure at the failure.

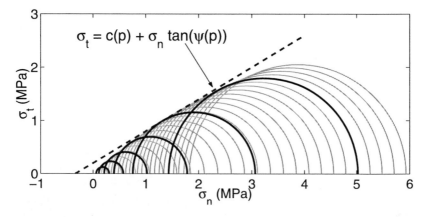

Figure 4.4: Mohr-Coulomb circles at the failure point for different pressures. The dotted line is tangent to the envelope curve of these circles.

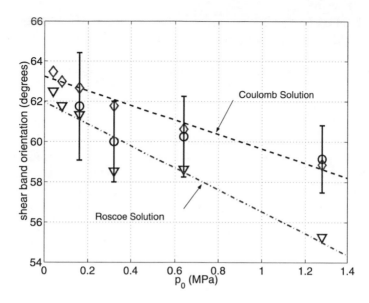

Figure 4.5: Shear band orientation (circles) compared to the Mohr-Coulomb solution (diamonds) and the Roscoe Solution (triangles). The lines correspond to linear fits.

of this nonlinearity is that the envelope of all Mohr-Coulomb circles at failure cannot be represented by a single straight line, as shown in Fig. 4.4. However, one can use the Mohr-Coulomb failure criterion in a *local sense*, by approaching the envelope around each Mohr-Coulomb circle by a straight line. This line can be constructed by taking the common tangent of the two circles at pressure $p - \Delta p$ and $p + \Delta p$. As shown in Fig.4.4, the resulting straight lines from these constructions lead to a dependence of the angle of friction and cohesive parameters with the pressure, so that they cannot be considered as material parameters.

This local Mohr-Coulomb analysis seems to be relatively consistent with the shear band orientation. Above the confining pressure of $p_0 > 160kPa$, we observed localization of strain as the typical mode of failure. This is given by a narrow zone in the sample where the dilatancy, the rotation of the particles, and the sliding between the grains are particularly intense. The measure of the shear band orientation for different confining pressures is shown in Fig. 4.2. The bars represent the uncertainty in the measure of the shear band, which is estimated as $\Delta\theta = atan(\Delta w/\Delta l)$, where Δw and Δl are the width and the length of the shear band.

Most of the experimental data from biaxial tests on sand report that the shear band orientation lies between the Mohr-Coulomb solution $\theta_C = 45° + \varphi/2$ and the Roscoe Solution $\theta_R = 45° + \Psi/2$ [13]. The latter is defined by the angle of dilatancy $\Phi = asin(d\epsilon_V/d\gamma)$, being $d\epsilon_V$ and $d\epsilon_\gamma$ the increments of volumetric and deviatoric strains at the failure [31]. These limits are shown in Fig. 4.2. We observe that the inclination angles are between these two angles with a tendency towards the Mohr-Coulomb solution. The fact that these angles do not coincide is consistent with the non-associativity of the plastic deformation of soils. This feature will be studied in detail in Chapter 6.

4.3 Hardening

Although the Mohr-Coulomb criterion is a simple and elegant approach to failure problems, this theory provides a too crude description of the actual behavior of granular materials. In particular, the granular materials do not show a perfectly elastic behavior up to the failure condition, but rather develop plastic deformations as a precursor behavior. This process is known as *hardening* in the literature of soil mechanics [11].

The hardening is investigated here at the grain level, by the evaluation of the plastic deformation between the grains during the simulation. For each polygon, the plastic deformation between two loading stages is calculated as $\xi = \sum_c |\Delta x^c - \Delta x_t^e|$ where Δx^c is the tangential displacement at each contact and Δx_t^e is the elastic part of this displacement. The latter is calculated after Eq. (3.1).

Fig. 4.6 shows the distribution of plastic displacements in four different loading stages. Irreversible deformations are observed at the very beginning of the loading. The plastic deformation is approximately uniform for small loadings, and it presents a progressive localization during the loading process. At failure, the shear band is identified by a narrow zone where the sliding between the grains is more intense than on average. After failure,

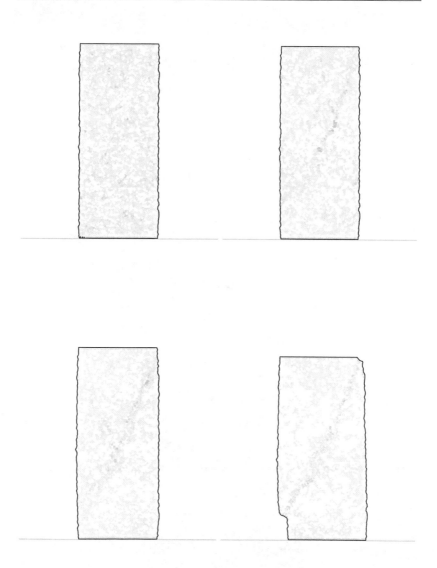

Figure 4.6: Plastic deformation at the grain during a loading of $\Delta\epsilon_1 = 0.001$. The intensity of the color represents the plastic deformation. The snapshot is taken for loading stages with $\epsilon_1 = 0.01, 0.02, 0.027$ (failure) and 0.07.

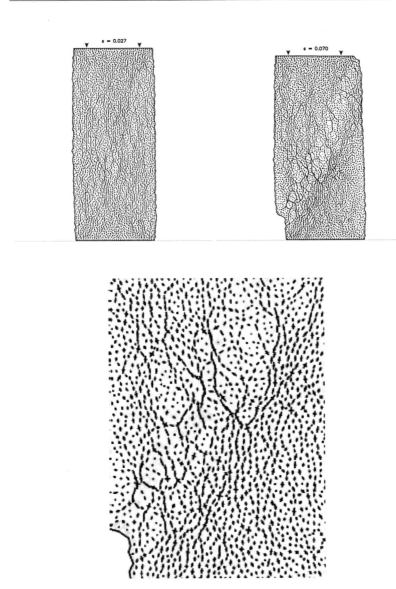

Figure 4.7: Principal stress directions of the individual grains at failure (left) and in the post-failure (right).

we observed two blocks moving one against another, separated by a shear band of some $6 - 8$ grains diameters.

This continuous hardening process has indeed lead many authors to the question of when the shear band occurs. It has been almost always assumed that the shear band occurs at peak stress or beyond the failure [13]. In contradiction to this, we observe some signals of localization of plastic deformations before the failure, as shown in part (b) of Fig. 4.6.

An explanation of this apparent contradiction can be found by looking at the distribution of the stress around the shear band. We calculate the average of the stress tensor at each particle as $\sigma_{ij} = \frac{1}{A} \sum_c f_i^c \ell_j^c$ where A is the area of the polygon, f_i^c is the contact force and ℓ_j^c is the branch vector, connecting the center of mass of the polygon with the center of application of the contact force. The sum goes over all the contacts of the particle. The principal stress direction at each grain is represented by a cross. The length of the lines represents how large the components are.

During loading, we observe that the principal stress direction goes almost perpendicular to the load direction, forming columnlike structures that are called chain forces. At failure, these chain forces start buckling , and the buckled chains gradually concentrate as shear bands in the post-failure process, which cause a growth of void ratio, and therefore a reduction of the strength in the shear band. For large deformations, one can see that the chain forces are perpendicular to the loading direction outside of the shear band, and they go almost perpendicular through the shear band. Due to this fact, there is an abrupt change of the stress in the parallel direction to the shear band, in agreement with the bifurcation analysis [13].

It is important to remark that the criterion to identify the moment of the arising of a shear band is still not well defined in our simulations. If we use the localization of sliding contacts, one may say that it appears before failure. On the other hand, if one uses the increase of the void ratio, or the buckling of the chain forces, it seems to appear in the post-failure behavior.

4.4 Critical states

The previous simulations were performed using perfect packings of polygons, with no porosity at the beginning of the simulation. This ideal case contrasts with realistic soils, where only porosities up to a certain value can be achieved. In this section, we present a method to create stable, irregular packings of polygons with a given porosity.

The porosity can be defined using the concept of void ratio. This is defined as the ratio of the volume of the void space to the volume of the solid material. It can be calculated as:

$$v = \frac{V_t}{V_f - V_0} - 1 \qquad (4.6)$$

This is given in terms of the area enclosed by the floppy boundary V_t, the

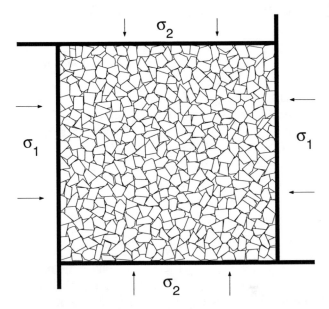

Figure 4.8: Polygonal assembly confined by a rectangular frame of walls.

sum of the area of the polygons V_f and the sum of the overlapping area between them V_0.

Of course, there is a maximal void ratio that can be achieved, because it is impossible to pack particles with an arbitrarily high porosity. The maximal void ratio v_m can be detected as follows. First, we move the walls until a certain void ratio is reached. Then, we find a critical value, above which the particles can be arranged without touching. Since there is no contacts, the assembly cannot support a load, and even applying gravity will cause it to compactify. For a void ratio below this critical value, there will be particle overlaps, and the assembly is able to sustain a certain load. This critical value is around 0.28.

The void ratio can be decreased by reducing the volume between the walls. The drawback of this method is that it leads to significant differences between the inner and outer parts of the boundary assembly, and it yields unrealistic overlaps between the particles, giving rise to enormous pressures. Alternatively, one can confine the polygons by applying gravity to the particles and on the walls. Particularly, homogeneous, isotropic assemblies can be generated by a gravitational field $\mathbf{g} = -g\mathbf{r}$ where $\mathbf{r}$ is the vector connecting the center of mass of the assembly with the center of the polygon.

When the sample is consolidated, repeated shear stress cycles are applied from the walls, superimposed to a confining pressure. The external load is imposed by applying a force $[p_c + q_c \sin(2\pi t/t_0)]W$ and $[p_c + q_c \cos(2\pi t/t_0)]H$ on the horizontal and vertical walls, respectively. W and H are the width and the height of the sample. If we take $p_c = 16kPa$ and $q_c < 0.4p_c$, we observe that the void ratio decreases as the number of cycles increases. Void ratios around 0.15 can be obtained. It is worth mentioning that after some thousands of cycles the void ratio is still slowly decreasing, making it difficult to identify this minimal value.

A third critical value for the void ratio can be obtained in the limit case of the biaxial deformations. When the polygonal samples are loaded, they pass through different configurations causing plastic deformations from the beginning of the loading. In the limiting case of large deformation, they reach a limit state where the void ratio and the stress fluctuate around con-

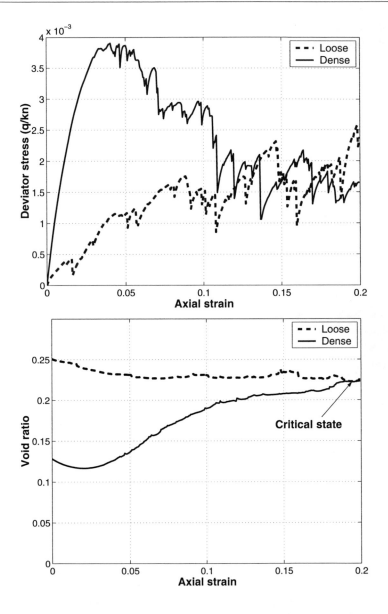

Figure 4.9: Stress-strain relation for dense and loose samples.

stant values under subsequent deformations. This state seems to be independent of the initial void ratio of the material, so that it can be considered as a *critical state* [32]. The existence of these states has been experimentally proven for clays and unbound granular soils [60] , and it has been the basis to develop new theoretical models which are known as *Critical State Models* [32].

The evolution of dense and loose samples to this critical state is shown in Fig. 4.9. When dense granular packings are loaded, it reaches a peak stress, and then the deviatoric stress decreases until it finally reaches a residual stress. Initially, the material compacts, and then dilates until the void ratio reaches a constant value that corresponds to its critical value. If the packing has a void ratio lower than the critical value, it deforms in such a way that there is not peak stress, and the void ratio increases until it reaches the critical value.

4.5   Concluding remarks

In order to perform a micromechanic investigation of the strain localization, numerical simulations on a discrete model with polygonal particles were performed. The results are summarized as follows:

- The onset of the plastic deformations proves to be the precursor mechanism of the shear band formation. We observe a progressive localization of plastic deformation before failure. After failure, a buckling of force chains is observed. This buckling leads to localized dilatancy and the onset of the shear band.

- The failure point shows a slight nonlinear dependence with the pressure. Then, the friction angle and cohesive factor of the Mohr-Coulomb analysis are not material parameters, because they depend on the stress state.

- The angle of orientation of the shear band lies in the range predicted by the bifurcation theories, with a tendency to be close to the corresponding angle of the Mohr-Coulomb analysis.

- The biaxial test for two extreme void ratios shows evidence for the existence of the critical states. A future investigation of the biaxial tests would require one to consider different initial densities in order to evaluate these states.

- The Mohr-Coulomb criterion gives a satisfactory description of the localized failure, but it provides an oversimplified description of the stress-strain relation. Plastic deformation is observed from the beginning of the load, ruling out an elastic regime. The dilatancy is observed before failure. This is an important ingredient in the failure analysis that is not taken into account in this theory.

Chapter 5

Incremental stress-strain relation

For many years the study of the mechanical behavior of soils was developed in the framework of linear elasticity [61] and the Mohr-Coulomb failure criterion [11]. However, since the boom of the developments of the nonlinear constitutive relations in 1968 [32], a great variety of constitutive models describing different aspects of soils have been proposed [1]. A crucial question has been brought forward: What it the most appropriate constitutive model to interpret the experimental result, or to implement a finite element code? Or more precisely, why is the constitutive relation I am using better than that one of the fellow next lab?

In the last years, the discrete element approach has been used as a tool to investigate the mechanical response of soils at the grain level [37]. Several average procedures have been proposed to define the stress [2, 46] and the strain tensor [6, 43] in terms of the contact forces and displacements at the individual grains. These methods have been used to perform a direct calculation of the incremental stress-strain relation of assemblies of disks [40] and spheres [48], without any a-priori hypothesis about the constitutive relation. Since these simple spherical geometries of the grains overestimate the role of rotations in realistic soils [49], it is interesting to see the incremental response using arbitrarily shaped particles.

In this chapter we investigate the incremental response in the quasistatic deformation of dense assemblies of polygonal particles by averaging the stress and strain tensors over a *representative volume element* of the sample. The strain envelope response is calculated in order to classify the incremental response of the discrete model. In Sec. 5.1 we present mi-cromechanical expressions for the average of the stress and strain tensors over a representative volume element. In Sec. 5.2 a short review of the incremental stress-strain is presented. The basic question of the validity incremental non-linearity of granular materials is discussed in basis of our numerical calculations. Finally, the calculation of the strain envelope response is presented in Sec. 5.3.

5.1 Homogenization

The aim of this section is to calculate the macromechanical quantities, the stress and strain tensors, from micromechanical variables of the granular assembly such as contact forces, rotations and displacements of individual grains.

A particular feature of granular materials is that both the stress and the deformation gradient are very concentrated in small regions around the contacts between the grains, so that they vary strongly on short distances. The standard homogenization procedure smears out these fluctuations by averaging these quantities over a *representative volume element* (RVE). The diameter d of the RVE must be such that $\delta \ll d \ll D$, where δ is the characteristic diameter of the particles and D is the characteristic length of the continuous variables.

We use here this procedure to obtain the averages of the stress and the strain tensors over a RVE in granular materials, which will allow us to compare the molecular dynamics simulations to the constitutive theories. We regard stress and strain to be continuously distributed through the grains, but concentrated at the contacts. It is important to comment that this averaging procedure would not be appropriate to describe the structure of the chain forces or the shear bands because typical variation of the stress corresponds to few particle diameters. Different averaging procedures using coarse-grained functions [17], or cutting the space in slide-shaped areas [6, 62], can deal with the question of how one can perform averages, and at the same time maintain these features.

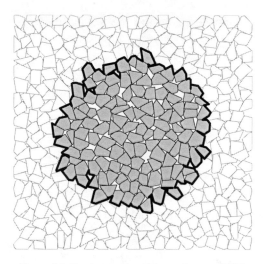

Figure 5.1: Representative volume element (RVE).

We will calculate the averages around a point x_0 of the granular sample taking a RVE calculated as follows: at the initial configuration, we select the grains whose center of mass are less than $d/2$ from x_0. Then the RVE is taken as the volume V enclosed by the initial configuration of the grains. See Fig. 5.1. The diameter d is taken, so that the averaged quantities are not sensible to the increase of the diameter by one particle diameter. does not affect the average stress more than 5%.

5.1.1 Micromechanical stress

The Cauchy stress tensor is defined using the force acting on an area element situated on or within the grains . Let f be the force applied on a surface element a whose normal unit vector is n. Then the stress is defined as the tensor satisfying [29]:

$$\sigma_{kj} n_k = lim_{a \to 0} f_j / a, \qquad (5.1)$$

where the Einstein summation convention is used. In absence of body forces, the equilibrium equations in every volume element lead to [61]:

$$\partial \sigma_{ij} / \partial x_i = 0. \tag{5.2}$$

We are going to calculate the average of the stress tensor $\bar{\sigma}$ over the RVE:

$$\bar{\sigma} = \frac{1}{V} \int_V \sigma dV \tag{5.3}$$

Since there is no stress at the voids of the granular media, the averaged stress can be written as the sum of integrals taken over the particles

$$\bar{\sigma} = \frac{1}{V} \sum_\alpha \int_{V_\alpha} \sigma_{ij} dV, \tag{5.4}$$

where V_α is the volume of the particle α. and N is the number of particles of the RVE. Using the identity

$$\frac{\partial(x_i \sigma_{kj})}{\partial x_k} = x_i \frac{\partial \sigma_{kj}}{\partial x_k} + \sigma_{ij}, \tag{5.5}$$

Eq. (5.2), and the Gauss theorem, Eq. (5.4) leads to

$$\bar{\sigma}_{ij} = \frac{1}{V} \sum_\alpha \int_{V_\alpha} \frac{\partial(x_i \sigma_{kj})}{\partial x_k} dV = \frac{1}{V} \sum_\alpha \int_{\partial V_\alpha} x_i \sigma_{kj} n_k da. \tag{5.6}$$

The right hand side is the sum over the surface integrals of each grain. ∂V_α represents the surface of the grain α and **n** is the unit vector perpendicular to the surface element da.

An important feature of granular materials is that the stress acting on each grain boundary is concentrated in the small regions near to the contact points. Then we can use the definition given in Eq. (5.1) to express this stress on particle α in terms of the contact force by introducing Dirac delta functions:

$$\sigma_{kj}n_k = \sum_{\beta=1}^{N_\alpha} f_j^{\alpha\beta}\delta(\mathbf{x} - \mathbf{x}^{\alpha\beta}),\tag{5.7}$$

where $\mathbf{x}^{\alpha\beta}$ and $\mathbf{f}^{\alpha\beta}$ are the position and the force at the contact β, and N_α is the number of contacts of the particle α. Replacing Eq. (5.7) into Eq. (5.6), we obtain

$$\bar{\sigma}_{ij} = \frac{1}{V}\sum_{\alpha\beta} x_i^{\alpha\beta} f_j^{\alpha\beta}.\tag{5.8}$$

Now we decompose $\mathbf{x}^{\alpha\beta} = \mathbf{x}^\alpha + \boldsymbol{\ell}^{\alpha\beta}$ where $\mathbf{x}^\alpha$ is the position of the center of mass and $\boldsymbol{\ell}^{\alpha\beta}$ is the branch vector, connecting the center of mass of the particle to the point of application of the contact force. Imposing this decomposition in Eq. (5.8), and using the equilibrium equations at each particle $\sum_\beta \mathbf{f}^{\alpha\beta} = 0$ we have

$$\bar{\sigma}_{ij} = \frac{1}{V}\sum_{\alpha\beta} \ell_i^{\alpha\beta} f_j^{\alpha\beta}.\tag{5.9}$$

From the equilibrium equations of the torques $\sum_\beta(\ell_i^{\alpha\beta} f_j^{\alpha\beta} - \ell_j^{\alpha\beta} f_i^{\alpha\beta}) = 0$ one obtains that this tensor is symmetric, i. e.,

$$\bar{\sigma}_{ij} - \bar{\sigma}_{ji} = 0.\tag{5.10}$$

This property leads to some simplifications, as we will see later.

5.1.2 Micromechanical strain

In elasticity theory, the strain tensor is defined as the symmetric part of the average of the displacement gradient with respect to the equilibrium configuration of the assembly. Using the law of conservation of energy, one can define the stress-strain relation in this theory [61]. In granular materials one cannot define the strain in this sense, because any loading involves a certain amount of plastic deformation at the contacts, so that it is not possible to define the initial reference state to calculate the strain. Nevertheless , one can define a strain tensor in the incremental sense. This is defined as the average of the displacement tensor taken from the deformation during the transition between two different stress states.

At the micromechanical level, the deformation of the granular materials is given by a displacement field $\mathbf{u} = \mathbf{r'} - \mathbf{r}$ at each point of the assembly. Here $\mathbf{r}$ and $\mathbf{r'}$ are the positions of a material point before and after deformation. The average of the strain and rotational tensors are defined as:

$$\bar{\epsilon} = \frac{1}{2}(F + F^T),\tag{5.11}$$

$$\bar{\omega} = \frac{1}{2}(F - F^T),\tag{5.12}$$

where F^T is the transpose of the deformation gradient F, which is defined as

$$F_{ij} = \frac{1}{V}\int_V \frac{\partial u_i}{\partial x_j}dV.\tag{5.13}$$

Using the Gauss theorem, we transform it into an integral over the surface of the RVE

$$F_{ij} = \frac{1}{V}\int_{\partial V} u_i n_j da,\tag{5.14}$$

where ∂V is the boundary of the volume element. We express this as the sum over the boundary particles of the RVE

$$F_{i,j} = \frac{1}{V}\sum_\alpha \int_{\partial V_\alpha} u_i n_j da,\tag{5.15}$$

where N_b is the number of boundary particles. To go further it is convenient to make some approximations. First, the displacements of the grains during deformation can be considered rigid except for the small deformations near to the contact that can be neglected. Then, if the displacements are small in comparison to the size of the particles, we can write the displacement of the material points inside of particle α as:

$$u_i(\mathbf{x}) \approx \Delta x_i^\alpha + e_{ijk}\Delta\phi_j^\alpha(x_k - x_k^\alpha), \qquad (5.16)$$

where $\Delta\mathbf{x}^\alpha$, $\Delta\phi^\alpha$ and $\mathbf{x}^\alpha$ are displacement, rotation and center of mass of the particle α which contains the material point $\mathbf{x}$, and e_{ijk} is the anti-symmetric unit tensor. Setting a parameterization for each surface of the boundary grains over the RVE, the deformation gradient can be explicitly calculated in terms of grain rotations and displacements by replacing Eq. (5.16) in Eq. (5.15).

In the particular case of a two-dimensional assembly of polygons, the boundary of the RVE is given by a graph $\{\mathbf{x}_1..\mathbf{x}_2, ..., \mathbf{x}_{N_b+1} = \mathbf{x}_1\}$ consisting of all the edges belonging to the external contour of the RVE, as shown in Fig. 5.1. In this case, Eq. (5.15) can be transformed as a sum of integrals over the segments of this contour.

$$F_{ij} = \frac{1}{V}\sum_{\beta=1}^{N_b}\int_{x_\beta}^{x_{\beta+1}} [\Delta x_i^\beta + e_{ik}\Delta\phi^\beta(x_k - x_k^\beta)]n_j^\beta ds, \qquad (5.17)$$

where $e_{ik} \equiv e_{i3k}$ and the unit vector $\mathbf{n}^\beta$ is perpendicular to the segment $\overrightarrow{x^\beta x^{\beta+1}}$. Here β is the index of this boundary segment; and $\Delta\mathbf{x}^\beta$, $\Delta\phi^\beta$ and $\mathbf{x}^\beta$ displacement, rotation and center of mass of the particle which contains this segment. Finally, by using the parameterization $\mathbf{x} = \mathbf{x}^\beta + s(\mathbf{x}^{\beta+1} - \mathbf{x}^\beta)$, where $(0 < s < 1)$, we can integrate Eq. (5.17) to obtain

$$F_{ij} = \frac{1}{V}\sum_\beta (\Delta x_i^\alpha + e_{ik}\Delta\phi^\alpha \ell_k^\beta)N_j^\beta, \qquad (5.18)$$

where $N_j^\beta = e_{j,k}(x_k^{\beta+1} - x_k^\beta)$ and $\boldsymbol{\ell} = (\mathbf{x}^{\beta+1} - \mathbf{x}^\beta)/2 - \mathbf{x}^\alpha$. The stress tensor can be calculated taking the symmetric part of this tensor using Eq. (5.11). Contrary to the strain tensor calculated for spherical particles [4], the individual rotation of the particles appears in the calculation of this tensor. This is given by the fact that for non-spherical particles the branch vector is not perpendicular to the contact normal vector, so that $e_{ik}\ell_k^\beta N_j^\beta \neq 0$.

5.2 Incremental theory

Since the stress and the strain tensor are symmetric, it is advantageous to simplify the notation by defining these quantities as six-dimensional vectors:

$$
\tilde{\sigma} = \begin{bmatrix} \sigma_{11} \\ \sigma_{22} \\ \sigma_{33} \\ \sqrt{2}\sigma_{23} \\ \sqrt{2}\sigma_{31} \\ \sqrt{2}\sigma_{13} \end{bmatrix}, \quad and \quad \tilde{\epsilon} = \begin{bmatrix} \epsilon_{11} \\ \epsilon_{22} \\ \epsilon_{33} \\ \sqrt{2}\epsilon_{23} \\ \sqrt{2}\epsilon_{31} \\ \sqrt{2}\epsilon_{13} \end{bmatrix} \tag{5.19}
$$

The coefficient $\sqrt{2}$ allows us to preserve the metric in this transformation: $\tilde{\sigma}_k\tilde{\sigma}_k = \bar{\sigma}_{ij}\bar{\sigma}_{ij}$. The relation between these two vectors will be established in the general context of the rate independent incremental constitutive relations. We will focus on two particular theoretical developments: the hypoplastic theory and the elasto-plastic models. The similarities and differences of both formulations are presented in the framework of the incremental theory as follows.

5.2.1 General framework

In principle, the mechanical response of soils can be described by a functional dependence of the stress $\tilde{\sigma}(t)$ at time t on the strain history $\{\tilde{\epsilon}(t')\}_{0<t'<t}$. However, the mathematical description of this dependence

turns out to be very complicated due to the non-linearity and irreversible behavior of these materials. An incremental relation, relating the incremental stress $d\tilde{\sigma}(t) = \sigma'(t)dt$ to the incremental strain $d\tilde{\epsilon}(t) = \epsilon'(t)dt$ and some internal variables χ accounting for the deformation history, enables us to avoid these mathematical difficulties [15]. This incremental scheme is also useful to solve geotechnical problems, since the finite element codes require that the constitutive relation be expressed incrementally.

Due to the large number of existing incremental relations, the necessity of a unified theoretical framework has been pointed out as an essential necessity to classify all the existing models [63]. In general, the incremental stress is related to the incremental strain by the following function:

$$\mathcal{F}_\chi(d\tilde{\epsilon}, d\tilde{\sigma}, dt). \qquad (5.20)$$

Let's look at the special case where there is no rate dependence in the constitutive relation. This means that this relation is not influenced by the time required during any loading tests, as corresponds to the quasi-static approximation. In this case $\mathcal{F}$ is invariant with respect to dt, and Eq. (5.20) can be reduced to:

$$d\tilde{\epsilon} = \mathcal{G}_\chi(d\tilde{\sigma}). \qquad (5.21)$$

In particular, the rate-independent condition implies that multiplying the loading time by a scalar λ does not affect the incremental stress-strain relation:

$$\forall \lambda, \quad \mathcal{G}_\chi(\lambda d\tilde{\sigma}) = \lambda \mathcal{G}_\chi(d\tilde{\sigma}). \qquad (5.22)$$

This equation means that $\mathcal{G}_\chi$ is an homogeneous function of degree one. In this case, the application of the Euler identity shows that Eq. (5.21) leads to

$$d\tilde{\epsilon} = M_\chi(\hat{\sigma})d\tilde{\sigma}, \qquad (5.23)$$

where $M_\chi = \partial \mathcal{G}_\chi / \partial(d\tilde{\sigma})$ and $\hat{\sigma}$ is the unitary vector defining the direction of the incremental stress

$$\hat{\sigma} = \frac{d\tilde{\sigma}}{|d\tilde{\sigma}|}. \tag{5.24}$$

Eq. (5.23) represents the general expression for the rate-independent constitutive relation. In order to determine the dependence of M on $\hat{\sigma}$, one can either perform experiments by taking different loading directions, or postulate explicit expressions based on a theoretical framework. The first approach will be considered in the next section, and the discussion about some existing theoretical models will be presented as follows.

5.2.2 Drucker-Prager models

The classical theory of elasto-plasticity has been established by Drucker and Prager in the context if metal plasticity [64]. Some extensions have been developed to describe soils, clays, rocks, concrete, etc. [11, 65]. Here, we present a short review of these developments in soil mechanics.

When a granular sample subjected to a confining pressure is loaded in the axial direction, it displays a typical stress-strain response as shown in the left part of Fig. 5.2. A continuous decrease of the stiffness (i.e. the slope of the stress-strain curve) is observed during loading. If the sample is unloaded, an abrupt increase in the stiffness is observed, leaving an irreversible deformation. One observes that if the stress is changed around some region below σ_A, which is called the *yield point*, the deformation is almost linear and reversible. The first postulate of the elasto-plastic theory establishes a stress region immediately below the yield point where only elastic deformations are possible.

Postulate 1: For each stage of loading there exists a finite region in the stress space where only reversible deformations are possible.

The simple Mohr-Coulomb model assumes a large elastic domain, so that the onset of plastic deformation occurs only at failure [11]. However, it has been experimentally shown that plastic deformation develops before

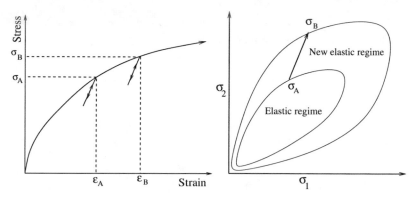

Figure 5.2: Evolution of the elastic regime **a)** stress-strain relation **b)** stress space.

failure [66]. In order to provide a consistent description of these experimental results with the elasto-plastic theory, it is necessary to suppose that the elastic domain changes with the deformation process [31]. This condition is schematically shown in Fig. 5.2. Let suppose that the sample is loaded until it reaches the stress σ_A and then it is slightly unloaded. If the sample is reloaded, it is able to recover the same stress-strain relation of the monotonic loading once it reaches the yield point σ_A again. If one increases the load to the stress σ_B, a new elastic response can be observed after a loading reversal. In the elasto-plasticity context, this result is interpreted by supposing that the elastic regime is expanded to a new stress region below the yield point σ_B.

Postulate 2: The elastic domain remains when the deformations take place inside it, and it changes as the plastic deformation evolves.

The transition from the elastic to the elasto-plastic response is extrapolated for more general deformations. Part (b) of Fig. 5.2 shows the evolution of the elastic region when the yield point moves in the stress space from σ_A to σ_B. The essential goal in the elasto-plastic theory is to find the correct description of the evolution of the elastic regime with the deformation, which is called the *hardening law*.

We will finally introduce the third basic assumption from elasto-plasticity theory:

Postulate 3: The strain can be separated in an elastic (recoverable) and a plastic (unrecoverable) component:

$$d\tilde{\epsilon} = d\tilde{\epsilon}^e + d\tilde{\epsilon}^p, \tag{5.25}$$

The incremental elastic strain is linked to the incremental stress by introducing an elastic tensor as

$$d\bar{\sigma} = D(\tilde{\sigma})d\tilde{\epsilon}^e. \tag{5.26}$$

To calculate the incremental plastic strain, we introduce a so-called *yield surface*, which encloses the elastic domain, as

$$f(\sigma, \kappa) = 0, \tag{5.27}$$

where κ is introduced as an internal variable, which describes the evolution of the elastic regime with the deformation. From experimental evidence, it has been shown that this variable can be taken as a function of the cumulative plastic strain [11, 65]

$$\epsilon^p \equiv \int_0^t \sqrt{d\epsilon_k d\epsilon_k} dt. \tag{5.28}$$

When the stress state reaches the yield surface, the plastic deformation evolves. This is assumed to be derived from a scalar function of the stress as follows:

$$d\epsilon_j^p = \Lambda \frac{\partial g}{\partial \sigma_j}, \tag{5.29}$$

where g is the so-called *plastic potential* function. following the Drucker-Prager postulates it can be shown that $g = f$ [64]. However, a considerable amount of experimental evidence has shown that in soils the plastic deformation is not perpendicular to the yield surfaces [67]. It is necessary to introduce this plastic potential to extend the Drucker-Prager models to the so-called *non-associated* models.

The parameter λ of Eq. (5.29) can be obtained from the so-called *consistence condition*. This condition comes from the second postulate, which establishes that after the movement of the stress state from $\tilde{\sigma}_A$ to $\tilde{\sigma}_B = \tilde{\sigma}_A + \tilde{d\sigma}$ the elastic regime must be expanded so that $df = 0$, as shown in Part (b) of Fig. 5.2. Using the chain rule one obtains:

$$df = \frac{\partial f}{\partial \sigma_i} d\sigma_i + \frac{\partial f}{\partial \kappa} \frac{\partial \kappa}{\partial \epsilon_j^p} d\epsilon_j^p = 0. \tag{5.30}$$

Replacing Eq. (5.29) in Eq. (5.30), we obtain the parameter Λ

$$\Lambda = -(\frac{\partial f}{\partial \kappa} \frac{\partial \kappa}{\partial \epsilon_j^p} \frac{\partial g}{\partial \sigma_j})^{-1} \frac{\partial f}{\partial \sigma_i} d\sigma_i. \tag{5.31}$$

We define the vectors $N_i^y = \partial f / \partial \sigma_i$ and $N_i^f = \partial g / \partial \sigma_i$ and the unit vectors $\hat{\phi} = \mathbf{N^y} / |\mathbf{N^y}|$ and $\hat{\psi} = \mathbf{N^f} / |\mathbf{N^f}|$ as the *flow direction* and the *yield direction*. The meaning of these vectors is explained below. In addition, the *plastic modulus* is defined as

$$h = -\frac{1}{|\mathbf{N^y}||\mathbf{N^f}|} \frac{\partial f}{\partial \kappa} \frac{\partial \kappa}{\partial \epsilon_j^p} \frac{\partial g}{\partial \sigma_j}. \tag{5.32}$$

Replacing Eq. (5.31) in Eq. (5.29) and using Eq. (5.32) we obtain:

$$d\tilde{\epsilon}^p = \frac{1}{h} \hat{\phi} \cdot d\tilde{\sigma} \, \hat{\psi}. \tag{5.33}$$

Note that this equation has been calculated for the case that the stress increment goes outside of the yield surface. If the stress increment takes place inside the yield surface, the second Drucker-Prager postulate establishes that $d\tilde{\epsilon}^p = 0$. Thus, the generalization of Eq. (5.33) is given by

$$d\tilde{\epsilon}^p = \frac{1}{h}\langle \hat{\phi} \cdot d\tilde{\sigma} \rangle \, \hat{\psi}, \tag{5.34}$$

where $\langle x \rangle = x$ when $x > 0$ and $\langle x \rangle = 0$ otherwise. This equation establishes a plastic deformation when the incremental stress has a component collinear with the yield direction, and an incremental plastic strain which points always to the flow direction. Finally, the total strain response can be obtained from Eqs. (5.25) and (5.34):

$$d\epsilon = D^{-1}(\sigma)d\sigma + \frac{1}{h}\langle \hat{\phi} \cdot d\tilde{\sigma} \rangle \, \hat{\psi}. \tag{5.35}$$

From this equation one can distinguish two zones in the incremental stress space where the incremental relation is linear. They are the so-called tensorial zones defined by Darve [15]. The existence of two tensorial zones, and the continuous transition of the incremental response at their boundary, are essential features of the elasto-plastic models.

Although the elasto-plastic theory has shown to work well for monotonically increasing loading, it has shown some deficiencies in the description of complex loading histories [68]. There is also an extensive body of experimental evidence that shows that the elastic regime is extremely small and that the transition from elastic to an elasto-plastic response is rather smooth.

The concept of *bounding surface* has been introduced to generalize the classical elasto-plastic concepts [69]. In this model, for any given state within the surface, a proper mapping rule associates a corresponding *image* stress point on this surface. A measure of the distance between the actual and the image stress points is used to specify the plastic modulus in terms of a plastic modulus at the image stress state. Besides the versatility of this theory to describe a wide range of materials, it has the advantage that the elastic regime can be considered as vanishingly small, and therefore used to give a realistic description of unbound granular soils.

It is the author's opinion that the most striking aspect of the bounding surface theory with vanishing elastic range is that it establishes a convergence point for two different approaches of constitutive modeling: the elasto-plastic approaches originated from the Drucker-Prager theory, and the more recently developed hypoplastic theories.

5.2.3   Hypoplastic models

In recent years, an alternative approach for modeling soil behavior has been proposed, which departs from the framework of the elasto-plastic theory [16, 24, 70]. The distinctive features of this approach are:

- **The absence of the decomposition of strain in plastic and elastic components.**

- **The statement of a nonlinear dependence of the incremental response with the strain rate directions.**

The most general expression has been provided by the so-called second order incremental non-linear models [24]. A particular class of these models which has received special attention in recent times is provided by the theory of hypoplasticity [16, 70]. A general outline of this theory was laid down by Kolymbas [16], leading to different formulations, such as the K-hypoplasticity developed in Karlsruhe [71, 72], and the CLoE-hypoplasticity originated in Grenoble [70]. In the hypoplasticity, the general constitutive equation takes the following form:

$$d\tilde{\sigma} = L(\tilde{\sigma}, v)d\tilde{\epsilon} + \tilde{N}(\tilde{\sigma}, v)|d\tilde{\epsilon}|. \qquad (5.36)$$

Where L is a second order tensor and $\tilde{N}$ is a vector, both depending on the current state of the material, the stress $\tilde{\sigma}$ and the void ratio v. Hypoplastic equations provide a simplified description of loose and dense unbound granular materials. A reduced number of parameters are introduced, which are very easy to calibrate [36].

In the theory of hypoplasticity, the stress-strain relation is established by means of an incremental nonlinear relation without any recourse to yield or boundary surfaces. This nonlinearity is reflected in the fact that the

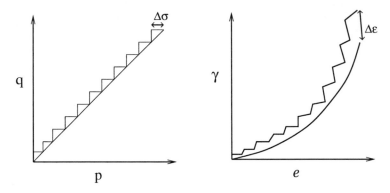

Figure 5.3: Smooth and stairlike stress paths and corresponding strain responses

relation between the incremental stress and the incremental strain given in Eq. (5.36) is always nonlinear. The incremental nonlinearity of the granular materials is still under discussion. Indeed, an important feature of the incremental nonlinear constitutive models is that they break away from the superposition principle:

$$d\tilde{\sigma}(d\tilde{\epsilon}_1 + d\tilde{\epsilon}_2) \neq d\tilde{\sigma}(d\tilde{\epsilon}_1) + d\tilde{\sigma}(d\tilde{\epsilon}_2), \tag{5.37}$$

which is equivalent to:

$$d\tilde{\epsilon}(d\tilde{\sigma}_1 + d\tilde{\sigma}_2) \neq d\tilde{\epsilon}(d\tilde{\sigma}_1) + d\tilde{\epsilon}(d\tilde{\sigma}_2) \tag{5.38}$$

An important consequence of this feature is addressed by Kolymbas [73] and Darve [24]. They consider two stress paths; the first one is smooth (proportional loading) and the second one results from the superposition of small deviations as shown in Fig. 5.3. The superposition principle establishes that the strain response of the stairlike path converges to the response of the proportional loading in the limit of arbitrarily small deviations. More precisely, the strain deviations $\Delta\epsilon$ and the steps of the stress increments $\Delta\sigma$ satisfy $\lim_{\Delta\sigma\to 0} \Delta\epsilon = 0$. For the hypoplastic equations, and in general for the incremental nonlinear models, this condition is never satisfied. For incremental relations with tensorial zones, this principle is satisfied whenever the increments take place inside one of these tensorial zones. It should be added that there is no experimental evidence disproving or confirming this rather questionable superposition principle.

In order to explore the validity of the superposition principle some numerical simulations were performed. Five different polygonal assemblies of 400 particles were used in the calculations. The stress was controlled in the RVE by applying a force $\mathbf{f}^\beta = -(p+q)\Delta x_2^\beta \hat{x}_1 + (p-q)\Delta x_1^\beta \hat{x}_2$ on each selected segment $\mathbf{T}^\beta = \Delta x_1^\beta \hat{x}_1 + \Delta x \beta_2 \hat{x}_2$ of the external contour of the assembly, where $\hat{x}_1$ and $\hat{x}_2$ are the unit vectors of the Cartesian coordinate system. The initial void ratio is around $\nu = 0.15$.

The components of the stress are reduced by $p = (\sigma_1 + \sigma_2)/2$ and $q = (\sigma_1 - \sigma_2)/2$, where σ_1 and σ_2 are the eigenvalues of the stress tensor. Equivalently, the coordinates of the strain are given by the sum $\gamma = \epsilon_1 + \epsilon_2$ and the difference $e = \epsilon_1 - \epsilon_2$ of the eigenvalues of the strain tensor. We use the convention that $e > 0$ means compression of the sample.

The part (a) of Fig. 5.4 shows the loading path during the proportional loading under constant lateral pressure. This path is also decomposed into pieces divided into two parts: one is an incremental isotropic loading ($\Delta p = \Delta \sigma$ and $\Delta q = 0$), the other is an incremental pure shear loading ($\Delta q = \Delta \sigma$ and $\Delta p = 0$). The length of the steps $\Delta \sigma$ varies from to $0.4p_0$ to $0.001p_0$, where $p_0 = 640 kPa$. The part (b) of Fig. 5.4 shows that as the steps decrease, the strain response converges to the response of the proportional loading. In order to verify this convergence, we calculate the difference between the strain response of the stairlike path $\gamma(e)$ and the proportional loading path $\gamma_0(e)$ as:

$$\Delta \epsilon = \max_e |\gamma(e) - \gamma_0(e)|, \tag{5.39}$$

for different steps sizes. This is shown in Fig. 5.5 for seven different polygonal assemblies. The linear interpolation of this data intersects the vertical axis at 3×10^{-7}. Since this value is below the error given by the quasistatic approximation, which is 3.13×10^{-4}, the results suggest that the responses converge to that one of the proportional load. Therefore we find that within our error bars the superposition principle is valid.

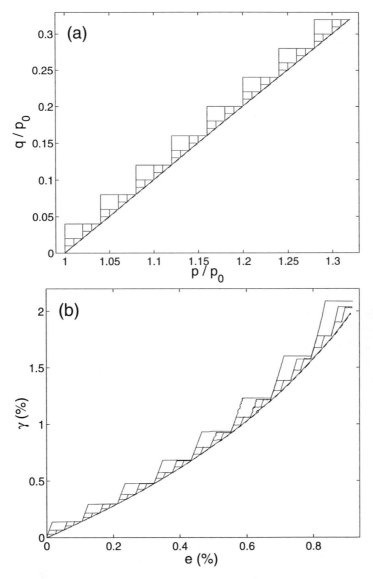

Figure 5.4: Comparison between strain responses obtained from MD simulations of a rectilinear proportional loading (with constant lateral pressure) and stairlike paths.

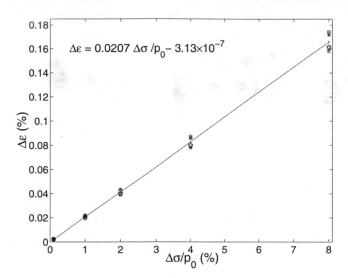

Figure 5.5: Distance between the response of the stairlike path and the proportional path.

A close inspection of the incremental response will show that the validity of the superposition principle is linked to the existence of tensorial zones in the incremental stress space. Before this, a short introduction to the strain envelope responses follows.

5.3 Incremental response

A graphical illustration of the particular features of the constitutive models can be given by employing the so-called *response envelopes*. They were introduced by Gudehus [63] as a useful tool to visualize the properties of a given incremental constitutive equation. A strain envelope response is defined as the image $\{d\tilde{\varepsilon} = \mathcal{G}(d\tilde{\sigma}, \tilde{\sigma})\}$ in the strain space of the unit sphere in the stress space, which becomes a potatolike surface in the stress space.

In practice, the determination of the strain envelope responses is difficult because it requires one to prepare many samples with identical material properties. Numerical simulations result as an alternative to the solution of

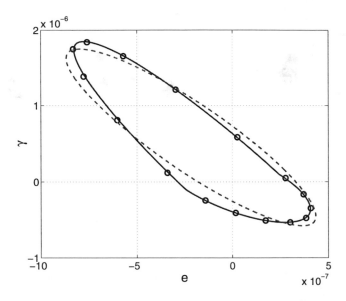

Figure 5.6: Numerical calculation of the incremental strain response. The dots are the numerical results. The solid curve represents the fit to the elasto-plastic theory. The dashed curve is the hypoplastic fit.

this problem. They allow one to create clones of the same sample, and to perform different loading histories in each one of them.

In recent years different discrete element methods have been used to calculate strain envelope responses. Disks [40] and spheres [47, 48] have been used in this calculation. Some of the results lead to the conclusion that the non-associated theory of elasto-plasticity is sufficient to describe the observed incremental behavior [40]. However, some recent investigations using three-dimensional loading paths of complex loading histories seem to contradict these results [48, 49].

In the case of a plane strain tests, where there is no deformation in one of the spatial directions, the strain envelope response can be represented in a plane. According to Eq. (5.36), This response results in a rotated, translated ellipse in the hypoplastic theory, whereas it is given by a continuous curve consisting of two pieces of ellipses in the elasto-plastic theory, as result from Eq. (5.35). It is then of obvious interest to compare these predictions with the stress envelope response of the experimental tests.

Fig. 5.6 shows the typical strain response resulting from the different stress controlled loading in a dense packing of polygons. Each point is obtained from the strain response in a specific direction of the stress space, with the same stress amplitude of $10^{-4}p_0$. We take $q_0 = 0.45p_0$ and $p_0 = 160kPa$ In this calculation. The best fit of these results with the envelopes response of the elasto-plasticity (two pieces of ellipses) and the hypoplasticity (one ellipse) is also shown in Fig. 5.6.

From these results we conclude that the elasto-plastic theory is more accurate in describing the incremental response of our model. One can draw to the same conclusion taking different initial stress values [21]. These results have shown that the incremental response can be accurately described using the elasto-plastic relation of Eq. (5.35). The validity of this equation is supported by the existence of a well defined flow rule for each stress state.

5.4   Concluding remarks

In this chapter we have obtained explicit expressions for the averaged stress and strain tensors over a RVE, in terms of the micromechanical variables, contact forces and the individual displacements and rotations of the grains.

The stress-strain relation on the RVE has been investigated by performing strain increments taking different directions in the stress space. The resulting incremental response has been compared to the elasto-plastic theory and the hypoplastic models. We found that the elasto-plastic theory, with two tensorial zones, provided a more accurate description of the incremental response than the hypoplastic theory.

Finally the principle of superposition has been investigated, with the aim to validate the existence of the tensorial zones of the incremental response. In contradiction to the incremental nonlinear models, the simulation results show that this principle is accurately satisfied.

In the next chapter we will separate the incremental response in an elastic

and a plastic contribution. We will see that a linear incremental elastic
response and a simple flow rule of plasticity gives a satisfactory description
of the mechanical behavior of this model.

Chapter 6

Analysis of the elasto-plastic response

In the previous chapter we showed that the stress envelope response of the polygonal packings fits better to the elasto-plastic models than the incremental non-linear models. We will see that from the calculation of the strain envelopes for different initial stress states, one can obtain the incremental response without establishing an elasto-plastic model a priori.

In this chapter we calculate the elasto-plastic response of a dense packing of polygons. From the analysis of the incremental response, we will show that the principal features of the deformation of soils can be reproduced by this simple model. In particular, the anisotropy of the stiffness tensor, the non-associated plastic flow rule, and the existence of failure modes inside the plastic limit surface will be discussed in the framework of the elasto-plastic theory. We will also discuss the relation of the constitutive models with the micromechanical arrangements, such as open and sliding contacts.

6.1   Introduction

6.2   Calculation of the incremental response

The elasto-plastic response of a perfect packing of polygons is calculated here by using molecular dynamics simulations. The stress is controlled from the boundary of assembles of 10×10 particles using the floppy boundary method from Sec. 3.3. The response of the assembly is investigated by defining the incremental stress and strain vector in a RVE of radius 8 particles from the center of the assembly. This calculation is performed in order to exclude the boundary effects from the calculations.

6.2.1 Basic assumptions

The micromechanical expression of the stress tensor is given by Eq. (5.9). Due to the symmetry of this tensor, their principal eigenvalues are real. From the principal components σ_1 and σ_2 of the stress tensor, one can define the stress vector:

$$\tilde{\sigma} = \begin{bmatrix} p \\ q \end{bmatrix} = \frac{1}{2} \begin{bmatrix} \sigma_1 + \sigma_2 \\ \sigma_1 - \sigma_2 \end{bmatrix}, \tag{6.1}$$

where p and q are the pressure and the deviatoric stress. The domain of admissible stresses is bounded by the failure surface. When the system reaches this surface, it becomes unstable and fails.

Before failure, the constitutive behavior can be obtained by performing small changes in the stress and evaluating the deformation response. An infinitesimal change of the stress vector $d\tilde{\sigma}$ produces an infinitesimal deformation of the RVE, which can be described by the incremental strain tensor. In Subsect. 5.1.2 this tensor was calculated from the average of the displacement gradient over the area of the RVE. It has been shown that it can be transformed in a sum over the boundary of the RVE.

$$d\epsilon_{ij} = \frac{1}{2A} \sum_b (\Delta u_i^b N_j^b + \Delta u_j^b N_i^b). \tag{6.2}$$

Here Δu^b is the displacement of the boundary segment, that is calculated from the linear displacement Δx and the angular rotation $\Delta \phi$ of the polygons belonging to it, according to Eq. (5.16). From the principal eigenvalues $d\epsilon_1$ and $d\epsilon_2$ of the symmetric part of this tensor, one can define the incremental strain vector as:

$$d\tilde{\epsilon} = \begin{bmatrix} de \\ d\gamma \end{bmatrix} = - \begin{bmatrix} d\epsilon_1 + d\epsilon_2 \\ d\epsilon_1 - d\epsilon_2 \end{bmatrix}. \tag{6.3}$$

By convention $de > 0$ corresponds to a compression of the sample. We are going to assume a rate-independent constitutive relation between the incremental stress and incremental strain tensor. According to Sec. 5.2, this can generally be written as:

$$d\tilde{\epsilon} = M(\hat{\theta}, \tilde{\sigma})d\tilde{\sigma}, \tag{6.4}$$

where $\hat{\theta}$ is the unitary vector defining a specific direction in the stress space:

$$\hat{\theta} = \frac{d\tilde{\sigma}}{|d\tilde{\sigma}|} \equiv \begin{bmatrix} \cos\theta \\ \sin\theta \end{bmatrix}, \quad |d\tilde{\sigma}| = \sqrt{dp^2 + dq^2}. \tag{6.5}$$

The constitutive relation results from the calculation of $d\tilde{\epsilon}(\theta)$, where each value of θ is related to a particular mode of loading. Some special modes are listed in Table 6.2.1.

In order to compare the resulting incremental response to the elasto-plastic theory, it is necessary to assume that the incremental strain can be separated into an elastic (recoverable) and a plastic (unrecoverable) component:

$$d\tilde{\epsilon} = d\tilde{\epsilon}^e + d\tilde{\epsilon}^p, \tag{6.6}$$

$0°$	isotropic compression	$dp > 0$	$dq = 0$
$45°$	axial loading	$d\sigma_1 > 0$	$d\sigma_2 = 0$
$90°$	pure shear	$dp = 0$	$dq > 0$
$135°$	lateral loading	$d\sigma_1 = 0$	$d\sigma_2 > 0$
$180°$	isotropic expansion	$dp < 0$	$dq = 0$
$225°$	axial stretching	$d\sigma_1 < 0$	$d\sigma_2 = 0$
$270°$	pure shear	$dp = 0$	$dq < 0$
$315°$	lateral stretching	$d\sigma_1 = 0$	$d\sigma_2 < 0$

Table 6.1: Principal modes of loading according to the orientation of $\hat{\theta}$

$$d\tilde{\epsilon}^e = D^{-1}(\tilde{\sigma})d\tilde{\sigma}, \tag{6.7}$$

$$d\tilde{\epsilon}^p = J(\theta, \tilde{\sigma})d\tilde{\sigma}. \tag{6.8}$$

Here, D^{-1} is the inverse of the stiffness tensor D, and $J = M - D^{-1}$ the flow rule of plasticity, which results from the calculation of $d\tilde{\epsilon}^e(\theta)$ and $d\tilde{\epsilon}^p(\theta)$.

6.2.2   The method

The method presented here to calculate the strain response has been used on experimental tests on sand [67]. It was introduced by Bardet [40] in the calculation of the incremental response using discrete element methods. This method will be used to determine the elastic $d\tilde{\epsilon}^e$ and plastic $d\tilde{\epsilon}^p$ components of the strain as function of the stress state $\tilde{\sigma}$ and the stress direction $\hat{\theta}$. Fig. 6.1 shows the three steps of the procedure:

1) The sample is driven to the stress state $\tilde{\sigma}$. First, it is isotropically compressed until it reaches the stress value $\sigma_1 = \sigma_2 = p - q$. Next, it is subjected to axial loading in order to increase the axial stress σ_1 to $p + q$. When the stress state with pressure p and deviatoric stress q is reached, the sample is allowed to relax.

2) Loading the sample from $\tilde{\sigma}$ to $\tilde{\sigma} + d\tilde{\sigma}$, the strain increment $d\tilde{\epsilon}$ is obtained. This procedure is implemented on different clones of the same sample, choosing different stress directions in each one of them, according to Eq. (6.5).

3) The samples are unloaded until the original stress state $\tilde{\sigma}$ is reached. Then one finds a remaining strain $d\tilde{\epsilon}^p$ that corresponds to the plastic component of the incremental strain.

The modulus of the stress increments is fixed to $|d\tilde{\sigma}| = 10^{-4}p_0$, where $p_0 = 160kPa$. This increments is chosen small enough, so that the unloaded

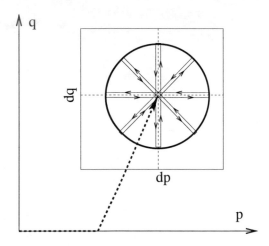

Figure 6.1: Procedure to obtain the constitutive behavior: 1) The sample is driven to the stress state $\tilde{\sigma}$, with pressure p and deviatoric stress q. 2) It is loaded from $\tilde{\sigma}$ to $\tilde{\sigma} + d\tilde{\sigma}$. 3) It is unloaded to the original stress state $\tilde{\sigma}$.

stress-strain path is approximately elastic. Thus, the difference $d\tilde{\epsilon}^e = d\tilde{\epsilon} - d\tilde{\epsilon}^p$ can be taken the elastic component of the strain.

This method is based on the assumption that the strain response after a reversal loading is completely elastic. However, numerical simulations have shown that this assumption is not strictly true, because sliding contacts are always observed during the unload path [27, 47]. In order to overcome this difficulty, Calvetti et al. [47] calculate the elastic part by removing the frictional condition from the algorithm setting $\mu = \infty$, and measuring the purely elastic response $\tilde{\epsilon}^{ns}$ of the assembly. Then the plastic component of the strain can be calculated as $d\tilde{\epsilon}^p = d\tilde{\epsilon} - d\tilde{\epsilon}^{ns}$.

In our simulation, we have observed that the plastic deformation during the reversal path is less than 1% of the corresponding value of the elastic response. Within this margin of error, the method proposed by Bardet can be taken as a reasonable approximation to describe the elasto-plastic response. It is worth mentioning that the plastic deformation after the loading reversal will result in a permanent accumulation of deformation when the sample is subjected to cyclic loading [27]. This topic will be discussed in Chapter 7.

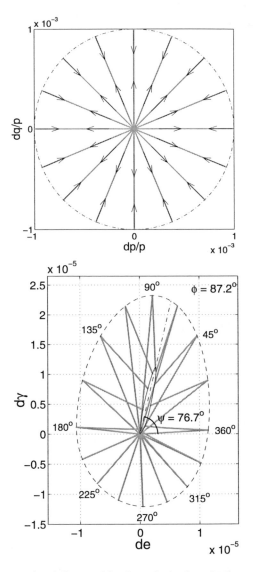

Figure 6.2: Stress - strain relation resulting from the load - unload test. Solid lines represent the paths in the stress and strain spaces. The dotted line gives the strain envelope response and the solid line is the plastic envelop response.

Fig. 6.2 shows the load-unload stress paths and the corresponding strain response when an initial stress state with $\sigma_1 = 200kPa$ and $\sigma_2 = 120kPa$ is chosen. The end of the load paths in the stress space maps into a strain envelope response $d\tilde{\varepsilon}(\theta)$ in the strain space. Likewise, the end of the unload paths maps into a plastic envelope response $d\tilde{\varepsilon}^p(\theta)$. The *yield direction* ϕ can be found from this response, as the direction in the stress space where the plastic response is maximal. In this example, this is around $\theta = 87.2°$. The *flow direction* ψ is given by the direction of the maximal plastic response in the strain space, which is around to $76.7°$. The fact that these directions do not agree reflects a *non-associated flow rule*, as it is observed in experiments on realistic soils [67]. We will explore this feature in the next section.

6.3 Constitutive relation

In this section, the elastic and plastic response envelopes are evaluated for different stress levels. The incremental stress-strain relation is calculated from the average of the envelope response over five different samples, each one with 10×10 particles. From the resulting incremental response, we examine the principal elements of the elasto-plastic theory: the elastic tensor, flow rule, failure surface, and the plastic limit surface.

6.3.1 Failure surface

The failure line was calculated by looking for the values of stress for which the system becomes unstable. For each pressure p, there is a critical deviatoric stress $q_c(p)$, below which the sample reaches a stable state with an exponential decay of its kinetic energy after the load is applied. For deviatoric stress values above the critical one, the sample becomes instable and fails. Fig. 6.3 shows the interface between these two stress states, which can be accurately fitted by the power law:

$$\frac{q}{q_c} = \left(\frac{p}{p_0}\right)^{\beta}.$$

(6.9)

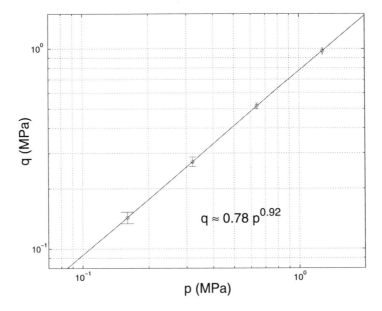

Figure 6.3: Failure limit. The continuous line represents the power law fit.

Here $p_0 = 1.0 MPa$ is the reference pressure, and $q_c = 0.78 \pm 0.03 MPa$. The power law dependence on the pressure, with exponent $\beta = 0.92 \pm 0.02$ implies a slight deviation from the Mohr-Coulomb theory. Empirical criteria of failure for most rocks [30] and soils [12] show a power law dependence of the form of Eq. (6.9).

6.3.2 Elastic tensor

Hooke's law of elasticity states that the stiffness tensor of isotropic materials can be written in terms of two material parameters, i.e. the Young modulus E and the Poisson ratio ν [61]. However, the isotropy is not fulfilled when the sample is subjected to deviatoric loading. Indeed, numerical simulations [2, 5] and photo-elastic experiments [74, 75] on granular materials show that the loading induces a significant deviation from isotropy in the contact network.

Anisotropy of the contact network

The anisotropy of the granular sample can be characterized by the distribution of the orientations of the branch vectors ℓ, as shown in Fig. 6.4, each branch vector connects the center of mass of the polygon to the center of application of the contact force. Fig. 6.4 shows the branch vectors for two different stages of loading. The structural changes of microcontacts are principally due to the opening of contacts whose branch vectors are nearly aligned around the direction perpendicular to the load. Let us call $\Omega(\varphi)\Delta\varphi$ the number of contacts per particle whose branch vector is oriented between the angles φ and $\varphi + \Delta\varphi$. The anisotropy of the contact distribution can be accurately described by a truncated series expansion.

$$\Omega(\varphi) \approx \frac{N_0}{2\pi}\big[a_0 + a_1 \cos(2\varphi) + a_2 \cos(4\varphi)\big]. \qquad (6.10)$$

Here $N = N_0 a_0$ is the average coordination number of the polygons, whose initial value $N_0 = 6$ reduces as the load is increased. The parameters a_0, a_1 and a_2 can be related respectively to the zero, second and fourth order fabric tensor defined in other works to characterize the contact distribution [5, 6, 76]. Here, they will be called *fabric coefficients*. The dependence of the fabric coefficients on the stress ratio q/p is shown in Fig. 6.5. We observe that for stress states satisfying $q < 0.4p$ there are almost no open contacts. Above this limit a significant number of contacts are open, leading to an anisotropy in the contact network. Fourth order terms in the Fourier expansion are necessary in order to accurately describe this distribution.

Anisotropic stiffness

We will investigate the effect of the anisotropy of the contact network on the stiffness of the material. The most general linear relation between the incremental stress and the incremental elastic strain for anisotropic materials is given by

$$d\sigma_{ij} = D_{ijkl}d\epsilon_{kl}^e \qquad (6.11)$$

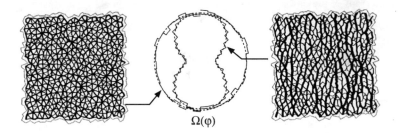

Figure 6.4: Distribution of branch vectors for $\sigma_1 = \sigma_2 = 160kPa$ (left) and $\sigma_1 = 272kPa$ and $\sigma_2 = 48kPa$ (right). The orientational distribution of branch vectors is shown for both cases.

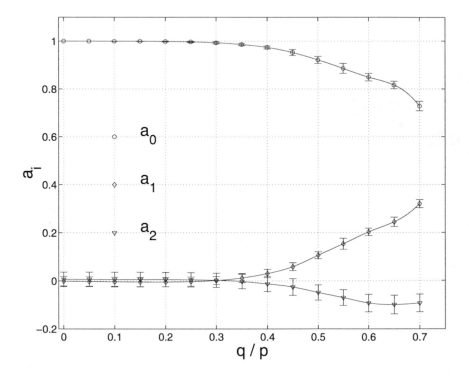

Figure 6.5: Fabric coefficients of the distribution of the contact normal vectors. They are defined in Eq. (6.10).

where D_{ijkl} is the stiffness tensor [61]. Since the stress and the strain are symmetric tensors, one can reduce their number of components from 9 to 6, and the number of components of the stiffness tensor from 81 to 36. Further, by transposing Eq. (6.11) one obtains that $D_{ijkl} = D_{jilk}$, which reduces the constants from 36 to 21. In the particular case of isotropic materials, it has been shown that the number of constants can be reduced to 2 [61]:

$$d\epsilon_{ij}^e = \frac{1}{E}[(1-\nu)d\sigma_{ij} - \nu\delta_{ij}d\sigma_{kk}]. \qquad (6.12)$$

Here E is the Young modulus and ν the Poisson ratio. The description of the general case of the anisotropic elasticity with 21 constants does not seem trivial. However, since we consider here only plane strain deformations, we can perform further simplifications. We take a coordinate system oriented in the principal stress-strain directions. Thus, the only nonzero components are $d\sigma_{11} \equiv d\sigma_1$ and $d\sigma_{22} \equiv d\sigma_2$ for the stress and $d\epsilon_{11} \equiv d\sigma_1$ and $d\epsilon_{22} \equiv d\sigma_2$ for the strain. The anisotropic elastic tensor connecting these components contains only three independent parameters. We can write Eq. (6.11) as

$$\begin{bmatrix} d\epsilon_1^e \\ d\epsilon_2^e \end{bmatrix} = \frac{1}{E} \begin{bmatrix} 1-\alpha & -\nu \\ -\nu & 1+\alpha \end{bmatrix} \begin{bmatrix} d\sigma_1 \\ d\sigma_2 \end{bmatrix}. \qquad (6.13)$$

The additional parameter α is included here to take into account the anisotropy. When $\alpha = 0$, we recover the Hooke's law of Eq. (6.12). Eq. (6.7) is calculated from Eq. (6.13) by performing the transformation in the coordinates of the volumetric strain $de = d\epsilon_1 + d\epsilon_2$ and deviatoric strain $d\gamma = d\epsilon_1 + d\epsilon_2$, and the pressure $p = (\sigma_1 + \sigma_2)/2$ and the deviatoric stress $q = (\sigma_1 - \sigma_2)/2$. One obtains:

$$\begin{bmatrix} de \\ d\gamma \end{bmatrix} = \frac{2}{E} \begin{bmatrix} 1-\nu & -\alpha \\ -\alpha & 1+\nu \end{bmatrix} \begin{bmatrix} dp \\ dq \end{bmatrix} \qquad (6.14)$$

In the isotropic case $\alpha = 0$ this matrix is diagonal. The inverse of the diagonal terms are the bulk modulus $K = E/2(1-\nu)$ and the shear mod-

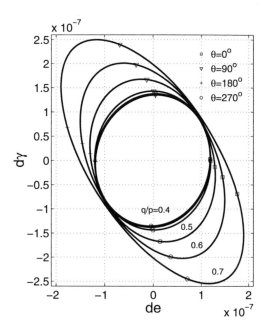

Figure 6.6: Elastic strain envelope responses $d\tilde{\epsilon}^e(\theta)$. They are calculated for a pressure $p = 160KPa$ and taking deviatoric stresses with $q = 0.0p$ (inner), $0.1p$, ...,$0.7p$ (outer).

ulus $G = E/2(1 + \nu)$. The anisotropy $\alpha \neq 0$ couples the compression mode with the shear deformation such that the compression of the sample will produce deviatoric deformation. This coupling can be observed from the inspection of the elastic part of the strain envelope responses $d\tilde{\epsilon}^e(\theta)$ as shown in Fig. 6.6. For stress values such as $q/p \leq 0.4$ the stress envelope responses collapse on to the same ellipse. This response can be described by Eq. (6.14) taking $\alpha = 0$. For stress values satisfying $q/p > 0.4$ there is a coupling between compression and shear deformations and it is necessary to take $\alpha \neq 0$ in Eq. (6.14).

Stiffness & Fabric

Comparing the calculation of the elastic response in Fig. 6.6 to the anisotropy of the contact network shown in Fig. 6.5, a certain correla-

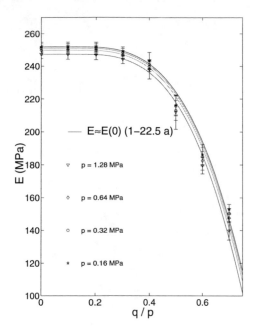

Figure 6.7: Young modulus. The solid line is the linear approximation of $E(a)$. See Eq. (6.22).

tion is evident between the stiffness tensor and the fabric coefficients of Eq. (6.10). We observe that Hooke's law is valid in the regime $q/p < 0.4$ where the contact network is isotropic. Moreover, we observe that the opening of the contacts, whose branch vectors are almost perpendicular to the direction of the load, produces a reduction of the stiffness in this direction. This results in an anisotropic elasticity.

We are going to find a simple relation between the orientational distribution of the contacts and the parameters of the stiffness. These three parameters are calculated from the elastic response by the introduction of the quadratic form of D^{-1}:

$$R(\theta) = \frac{dp\,de^e + dq\,d\gamma^e}{dp^2 + dq^2}. \tag{6.15}$$

This function can be directly obtained from the elastic part of the strain

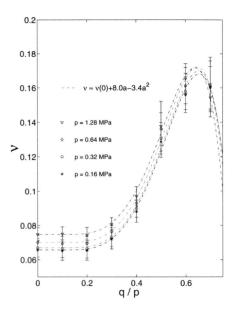

Figure 6.8: Poisson ratio. The dashed line is the quadratic approximation of $\nu(a)$. See Eq. (6.22).

envelope response $d\tilde{\epsilon}^e(\theta)$. On the other hand, replacing Eq. (6.14) in Eq. (6.15) one can express R in terms of the parameters of the stiffness tensor:

$$R(\theta) = \frac{2}{E}\big[1 - \nu\cos(2\theta) - \alpha\sin(2\theta)\big]. \tag{6.16}$$

Using this equation, the parameters E, ν and α are evaluated from the Fourier coefficients of R:

$$\frac{1}{E} = \frac{1}{4\pi}\int_0^{2\pi} R(\theta)d\theta, \tag{6.17}$$

$$\nu = -\frac{E}{2\pi}\int_0^{2\pi} R(\theta)\cos(2\theta)d\theta, \tag{6.18}$$

$$\alpha = -\frac{E}{2\pi}\int_0^{2\pi} R(\theta)\sin(2\theta)d\theta. \tag{6.19}$$

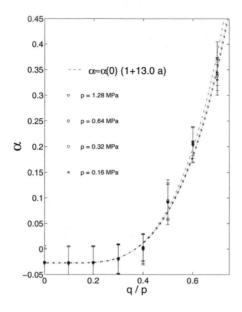

Figure 6.9: Anisotropy parameter. The dashed line is the linear approximation of $\alpha(a)$. See Eq. (6.22).

Figs. 6.7, 6.8 and 6.9 show the results of the calculation of the Young modulus E, the Poisson ratio ν and the anisotropy factor α. Below the limit of isotropy, Hooke's law can be applied: $E \approx E_0$, $\nu \approx \nu_0$ and $\alpha \approx 0$. On the other hand, above the limit of isotropy a reduction of the Young modulus is found, along with an increase of the Poisson ratio and the anisotropy factor. In order to evaluate the dependence of these parameters on the fabric coefficients a_i of Eq. (6.10), we introduce an internal variable measuring the degree of anisotropy. This variable is denoted by a and is defined as the percentage change of the total number of contacts.

$$a = \frac{N_0 - N}{N} \approx 1 - a_0 \tag{6.20}$$

where a_0 is defined in Eq. (6.10). The dependence of the parameters of the stiffness tensor on the internal variable a is evaluated by developing the Taylor series around $a = 0$:

$$\begin{aligned}
E(a) &= E(0) + E'(0)a + O\left(a^2\right), \\
\alpha(a) &= \alpha(0) + \alpha'(0)a + O\left(a^2\right), \\
\nu(a) &= \nu(0) + \nu'(0)a + \nu''(0)a^2 + O\left(a^3\right).
\end{aligned}$$

(6.21)

The coefficients of these expansions are calculated from the best fit of those expresions. Figs. 6.7 and 6.9 show that the linear approximation is good enough to reproduce the Young modulus and the anisotropy factor. The fit of the Poisson ratio, is shown in Fig. 6.8. Fitting with only one internal variable requires the inclusion of a quadratic approximation. To obtain more accurate relations, it may be necessary to introduce a more complex dependence with the fabric coefficients of Eq. (6.10).

6.3.3 Plastic deformations

In the elasto-plastic models of soils the plastic deformation is calculated by introducing a certain number of hypothetical surfaces [32, 65, 66, 77]. In the Drucker-Prager models, the so-called plastic flow rule is calculated from the yield surface and the plastic potential [32, 65, 66]. In the bounding surface plasticity, it is calculated from the loading surface and bounding surfaces [69, 77]. We will see that it is possible to calculate the relevant parameters of the flow rule of plasticity directly from the stress envelope response $d\bar{\varepsilon}^p(\theta)$ without introducing such abstract surfaces.

Flow rule

In Fig. 6.2 we found that the plastic envelope response lies almost on a straight line, as is predicted by the Drucker-Prager theory. This motivates us to define the parameters describing the plasticity in the same way as this theory: i.e. the yield direction ϕ, the flow direction ψ, and the plastic modulus h.

The yield direction is given by the incremental stress direction ϕ with maximal plastic deformation

$$|d\tilde{e}^{p}(\phi)| = \max_{\theta} |d\tilde{e}^{p}(\theta)|, \tag{6.22}$$

The flow direction is defined from the orientation of the plastic response at its maximum value

$$\psi = atan2(d\gamma^{p}, de^{p})\,|_{\theta=\phi} \tag{6.23}$$

Here $atan2(y, x)$ is the four quadrant inverse tangent of the real parts of the elements of x and y. ($-\pi <= atan2(y, x) <= \pi$). The plastic modulus is defined from the modulus of the maximal plastic response

$$\frac{1}{h} = \frac{|d\tilde{e}^{p}(\phi)|}{|d\tilde{\sigma}|}. \tag{6.24}$$

The incremental plastic response can be expressed in terms of these quantities. Let us define the unitary vectors $\hat{\psi}$ and $\hat{\psi}^{\perp}$. The first one is oriented in the direction of ψ and the second one is the rotation of $\hat{\psi}$ of 90°. The plastic strain is written as:

$$d\tilde{e}^{p}(\theta) = \frac{1}{h}\left[\kappa_{1}(\theta)\hat{\psi} + \kappa_{2}(\theta)\hat{\psi}^{\perp}\right], \tag{6.25}$$

where the plastic components $\kappa_{1}(\theta)$ and $\kappa_{2}(\theta)$ are given by

$$\begin{aligned} \kappa_{1}(\theta) &= h(d\tilde{e}^{p} \cdot \hat{\psi}) \\ \kappa_{2}(\theta) &= h(d\tilde{e}^{p} \cdot \hat{\psi}^{\perp}). \end{aligned} \tag{6.26}$$

These functions are calculated from the resulting plastic response taking different stress values. The results are shown in Fig. 6.10. We found that the functions $\kappa_{1}(\theta - \psi)$ collapse on to the same curve for all the stress states. This curve fits well to a cosine function, truncated to zero for the negative values. The profile κ_{2} depends on the stress ratio we take. This dependency is difficult to evaluate, because the values of this function are

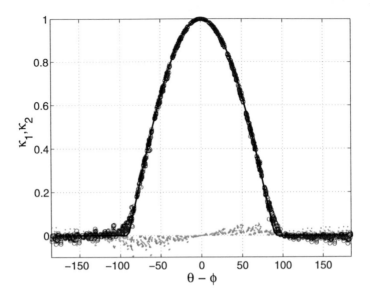

Figure 6.10: Plastic components $\kappa_1(\theta)$ (dots) and $\kappa_2(\theta)$ (pluses). The results for different stress values have been superposed. The solid line represents the truncated cosine function.

of the same order as the statistical fluctuations. In order to simplify the description of the plastic response, the following approximation is made:

$$\kappa_2(\theta) \ll \kappa_1(\theta) \approx \langle \cos(\theta - \phi) \rangle = \langle \hat{\phi} \cdot \hat{\theta} \rangle, \tag{6.27}$$

where $\langle \cdot \rangle$ defines the function

$$\langle x \rangle = \begin{cases} 0 & : \quad x \leq 0, \\ x & : \quad x > 0. \end{cases} \tag{6.28}$$

Now, the flow rule results from the substitution of Eqs.(6.25) and (6.27) into Eq. (6.8):

$$d\tilde{\epsilon}^p(\theta) = J(\theta)d\tilde{\sigma} = \frac{\langle \hat{\phi} \cdot d\tilde{\sigma} \rangle}{h} \hat{\psi}. \tag{6.29}$$

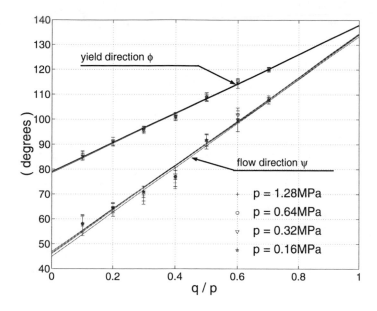

Figure 6.11: The flow direction and the yield direction of the plastic response. Solid lines represent a linear fit.

Although we have neither introduced yield functions nor plastic potentials, we recover the same structure of the plastic deformation obtained in Sec. 5.2 from the Drucker-Prager analysis. This result suggests the possibility to measure such surfaces directly from the envelope responses without need of an a-priori hypothesis about these surfaces. The next step is to verify the validity of the Drucker-Prager normality postulate, which states that the yield function must coincide with the plastic potential function [64].

Normality postulate

The Drucker normality postulate establishes that the flow direction is always perpendicular to the yield surface [64]. Since it was introduced to describe the plasticity in metals, the question naturally arises as to its validity for the plastic deformation for soils. With this aim, the yield direction and the flow direction have been calculated for different stress states. The results prove that both angles are quite different, as shown in Fig. 6.11. A large amount of experimental evidence has also indicated a clear deviation from Drucker's normality postulate [78].

It is not surprising that the Drucker postulate, which has been established for metal plasticity, is not fulfilled in the deformation of granular materials. Indeed, the principal mechanism of plasticity in granular materials is the rearrangement of the grains by the sliding contacts. This is not the case of microstructural changes in the metals, where there is no frictional resistance [79]. On the other hand, the sliding between the grains can be well handled in the discrete element formulation, which more adequately describes the soil deformation.

Yield function and plastic potential

The fact that the Drucker postulate is not fulfilled in the deformation of the granular materials has led to the so-called non-associated theory of plasticity [66]. This theory introduces a yield surface defining the yield directions and a plastic potential function, which defines the direction of the plastic strain.

Both, yield surfaces and plastic potential function can be calculated from the yield and flow direction, which in turn are calculated from the strain envelope response using Eqs. (6.22) and (6.23). According to Fig. 6.11, they follow approximately a linear dependence with the stress ratio q/p:

$$\begin{aligned}
\phi &= \phi_0 + \phi_0'\frac{q}{p}, \\
\psi &= \psi_0 + \psi_0'\frac{q}{p}.
\end{aligned} \tag{6.30}$$

The four parameters $\psi_0 = 46°\pm0.75°$, $\psi_0' = 88.3°\pm0.6°$, $\phi_0 = 78.9°\pm0.2°$ and $\phi_0' = 59.1° \pm 0.4°$ are obtained from a linear fit of the data. This linear dependence with the stress ratio has been shown to fit well with the experimental data in triaxial [10] and biaxial [80] tests on sand. In fact, this implies that the plastic potential function and the yield surfaces have the same shape, independent on the stress level. This is a basic assumption from the isotropic hardening models [65].

From Eq. (6.30), one can see that there is a transition from contractancy to dilatancy around $q/p = 0.5$. This transition is typically observed in dense sand under biaxial loading [65]. A consequence of this linear dependency is that $\psi \neq 0$ when $q = 0$. This implies the existence of deviatoric plastic strain when the sample is initially under isotropic loading conditions, which has been also predicted in the original Cam-Clay model [32].

The existence of deviatoric plastic deformation under extremely small loading appears to be in contradiction with the fact that the contact network remains isotropic below of a certain stress ratio (see Sec. 6.3.2). This matter has also been discussed by Nova [65], who introduced some modifications in the Cam-Clay model in order to satisfied the isotropic condition [65]. However, we are going to show that the orientational distribution of the sliding contacts departs from the isotropy for extremely small deviatoric loadings.

Plasticity & sliding contacts

Under small deformations, the individual grains of a realistic soil behave approximately rigidly, and the plastic deformation of the assembly is due principally to sliding contacts (eventually there is fragmentation of the grains, which is not going to be taken into account here). A complete understanding of soil plasticity is possible, in principle, by the investigation of the micromechanical arrangement between the grains. We present here some observations about the anisotropy induced by loading in the subnetwork of the sliding contacts. This investigation will be useful to understand some features of plastic deformation.

The sliding condition at the contacts is given by $|f_t| = \mu f_n$, where f_n and f_t are the normal and tangential components of the contact force, and μ is the friction coefficient. When the sample is isotropically compressed, we observe a significant number of contacts reaching the sliding condition. If the sample has not been previously sheared, the distribution of the orientation of the branch vectors of all the sliding contacts is isotropic.

This isotropy, however, is broken when the sample is subjected to the slightest deviatoric strain. In effect, at the very beginning of the loading,

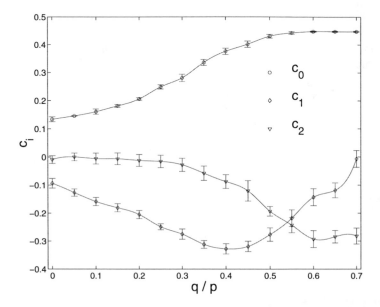

Figure 6.12: Fabric coefficients of the distribution of the contact normal vectors. They are defined in Eq. (6.31).

most of the sliding contacts whose branch vector is oriented nearly parallel to the loading direction leave the sliding condition. The anisotropy of the sliding contacts is investigated by introducing the polar function $\Omega^s(\varphi)$, where $\Omega^s(\varphi)\Delta\varphi$ is the number of sliding contacts per particle whose branch vector is oriented between φ and $\varphi + \Delta\varphi$. This can be approximated by a truncated Fourier expansion:

$$\Omega^s(\varphi) \approx \frac{N_0}{2\pi}\big[c_0 + c_1 \cos(2\varphi) + c_2 \cos(4\varphi)\big]. \qquad (6.31)$$

The coefficient of this expression are shown in Fig. 6.12. Fig. 6.13 shows the orientational distribution of sliding contacts for different stress ratios. For low stress ratios, the branch vectors ℓ of the sliding contacts are oriented nearly perpendicular to the loading direction. Sliding occurs perpendicular to ℓ, so in this case it must be nearly parallel to the loading direction. Then, the plastic deformation must be such as $d\epsilon_2^p \ll d\epsilon_1^p$, so Eq. (6.23) yields a flow direction of $\psi \approx 45°$, in agreement with Eq. (6.30).

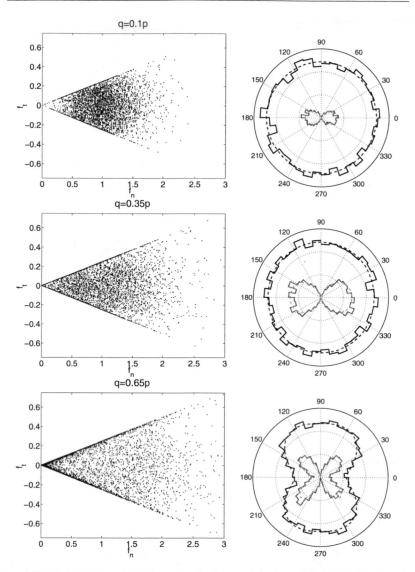

Figure 6.13: Left: force distribution for the stress ratios $q/p = 0.1, 0.35, 0.65$. Here f_t and f_n are the tangential and normal components of the force. They are normalized by the mean value of f_n. Right: orientational distribution of the contacts $\Omega(\varphi)$ (outer) and of the sliding contacts $\Omega^s(\varphi)$ (inner). φ represents the orientation of the branch vector.

Increasing the deviatoric strain results in an increase of the number of the sliding contacts. The average of the orientations of the branch vectors with respect to the load direction decreases with the stress ratio, which in turn results in a change of the orientation of the plastic flow. Close to the failure, some of the sliding contacts whose branch vectors are nearly parallel to the loading direction open, giving rise to a butterfly shape distribution, as shown in Fig. 6.13. In this case, the mean value of the orientation of the branch vector with respect to the loading direction is around $\varphi = 38°$, which means that the sliding between the grains occurs principally around $52°$ with respect to the vertical. This provides a crude estimate of the ratio between the principal components of the plastic deformation as $d\epsilon_2^p \approx -d\epsilon_1^p \tan(52°)$. According to Eq. (6.23) this gives an angle of dilatancy of $\psi = atan2(d\gamma^p, de^p) \approx 97°$. This crude approximation is reasonably close to the angle of dilatancy of $103.4°$ calculated from Eq. (6.30).

A fairly close correlation between the orientation of the sliding contacts and the angle of dilatancy has also been reported by Calvetti et al. [47] using molecular dynamic simulations in triaxial tests. This correlation suggests that the plastic deformation of soils can be micromechanically described by the introduction of the fabric constants c_i of the equation 6.31 in the constitutive relations. This investigation would lead to new structure tensors capturing the non-associativity of plastic deformation.

Plastic modulus

The plastic modulus h defined in Eq. (6.24) is related to the incremental plastic strain in the same way as the Young modulus is related to the incremental elastic strain. Thus, just as we related the Young modulus to the average coordination number of the polygons, it is reasonable to connect h to the fraction of sliding contacts $n_s = N^s/N$. Here N and N_s are the total number of contacts and the number of sliding contacts.

Fig. 6.14 shows the relation between the hardening and the fraction of the sliding contacts taken from $q = 0.0, 0.1p, ...0.07p$ with different pressures. The results can be fitted to an exponential relation

$$h = h_o \exp(-n_s/n_0) \qquad (6.32)$$

Where $h_0 = 80.0 \pm 0.4 GPa$ and $n_0 = 0.066 \pm 0.002$. This exponential dependence contrasts with the linear relation between the Young modulus and the number of contacts obtained in Sec. 6.3.2. From this comparison, it follows that when the number of contacts is such that $n_s > n_0$, the deformation is not homogeneous, but is principally concentrated more and more at the sliding contacts as their number increases.

The above results suggest that it is possible to establish a dependency of the flow rule on the anisotropy of the subnetwork of the sliding contacts. This relation is more appropriate than just an explicit relation between the flow rule and the stress, which probes to be only valid in the case of monotonic loading [10]. Since the stress can be expressed in terms of micromechanical variables, branch vectors and contact forces, the identification of those internal variables measuring anisotropy and force distribution would provide a more general description of the dependence of the flow rule on the history of the deformation.

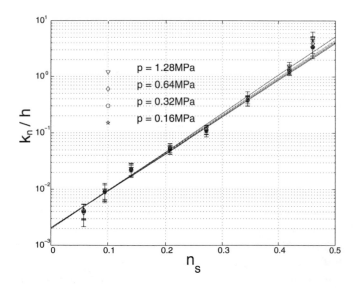

Figure 6.14: Inverse of hardening modulus versus fraction of sliding contacts n_s.

6.3.4 Yield function

In the previous section, the yield direction defining the flow rule of plasticity was calculated from the incremental strain response $d\tilde{\epsilon}^p(\theta)$. According to the Drucker-Prager theory, this direction must correspond to the perpendicular direction of the yield surface. This surface encloses a hypothetical region in the stress space where only elastic deformations are possible [64]. The determination of such a yield surface is essential to determine the dependence of the strain response on the history of the deformation.

We attempt to detect the yield surface by using a standard procedure proposed in experiments with sand [78]. Fig. 6.15 shows this procedure. Initially the sample is subjected to an isotropic pressure. Then the sample is loaded in the axial direction until it reaches the yield-stress state with pressure p and deviatoric stress q. Since plastic deformation is found at this stress value, the point (p, q) can be considered as a classical yield point. Then, the Drucker-Prager theory assumes the existence of a yield surface containing this point. In order to explore the yield surface, the sample is unloaded in the axial direction until it reaches the stress point with pressure $p - \Delta p$ and deviatoric stress $q - \Delta p$ inside the elastic regime. Then the yield surface is constructed by reloading in different directions in the stress space. In each direction, the new yield point must be detected by a sharp change of the slope in the stress-strain curve, indicating plastic deformations.

Fig. 6.16 shows the strain response taking different load directions in the same sample. The initial stress of the sample is given by $q_0 = 0.5p_0$ and $p_0 = 160kPa$. If the direction of the reload path is the same as that of the original load ($45°$), we observe a sharp decrease of stiffness when the load point reaches the initial yield point, which corresponds to the origin in Fig. 6.16. However, if one takes a direction of reloading different from $45°$, the decrease of the stiffness with the loading becomes smooth. Since there is no straightforward way to identify those points where the yielding begins, the yield function, as it was introduced by Drucker and Prager [64] in order to describe a sharp transition between the elastic and plastic regions, is not consistent with our results.

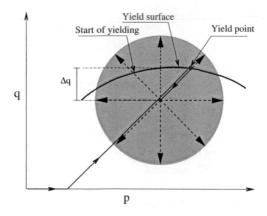

Figure 6.15: Method to obtain the yield surface. Load-unload-reload tests are performed taking different directions in the reload path. In each direction, the point of the reload path where the yielding begins is marked. The yield function is constructed by connecting these points.

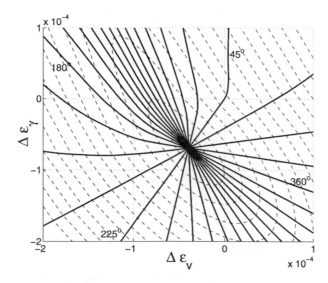

Figure 6.16: Strain responses according to Fig. 6.15. The solid lines show the strain response from different reload directions. The dashed contours represent the strain envelope responses for different stress increments $|\Delta\sigma|$.

6.4   Instabilities

Instability has been one of the classical topics of soil mechanics. Localization from a previously homogeneous deformation to a narrow zone of
intense shear is a common mode of failure of soils [12, 31, 66]. The Mohr-
Coulomb criterion is typically used to understand the principal features of
the localization. This criterion was improved by the Drucker condition,
based on the hypothesis of the normality, which results in a plastic flow
perpendicular to the yield surface [64]. This theory predicts that the instability appears when the stress of the sample reaches the plastic limit
surface. This surface is given by the stress states where the plastic deformation becomes infinite. Since the normality postulate is not fulfilled in
our calculations, it is interesting to see if the Drucker stability criterion is
still valid.

According to the flow rule from Eq. (6.29), the plastic limit surface can be
found by looking for the stress values where the plastic modulus vanishes.
First, we perform a suitable fitting of the dependence of the plastic modulus
on the stress. Fig. 6.17 shows that it can be fitted by the following power
law relation:

$$h = h_0 \left[1 - \frac{q}{q_0} (\frac{p_0}{p})^\vartheta \right]^\eta .$$

(6.33)

This is given in terms of the four parameters: The plastic modulus $h_0 =
14.5 \pm 0.05$ at $q = 0$, the constant $q_0 = 0.85 \pm 0.05$, and the exponents
$\eta = 2.7 \pm 0.04$ and $\vartheta = 0.99 \pm 0.02$. Then, the plastic limit surface is
given by the stress states with zero plastic modulus:

$$\frac{q_p}{q_0} = \left(\frac{p}{p_0} \right)^\vartheta .$$

(6.34)

We found that the failure surface, which is given in Eq. (6.9), does not
correspond to the plastic limit surface. By comparing both equations one
observes that during loading the instabilities appear before reaching the

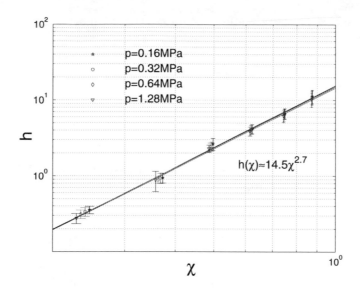

Figure 6.17: Plastic modulus. The solid line is a power law fit with respect to the variable $\chi = 1 - (p/p_0)^\vartheta q/q_0$.

plastic limit surface. Theoretical studies have also shown that in the case of non-associated materials, (i.e. where flow direction does not agree with the yield direction) the instabilities can appear strictly inside of the plastic limit surface [15]. In this context, the question of instability must be reconsidered beyond the Drucker condition.

The stability for non-associated elasto-plastic materials has been investigated by Hill, who established the following sufficient stability criterion [79].

$$\forall d\tilde{\varepsilon}, \quad d\tilde{\sigma} \cdot d\tilde{\varepsilon} > 0. \tag{6.35}$$

The analysis of this criterion of stability will be presented here based on the constitutive relation obtained in the last section:

$$d\tilde{\varepsilon} = D^{-1}d\tilde{\sigma} + \frac{\langle \hat{\phi}^T d\tilde{\sigma} \rangle}{h}\hat{\psi} \tag{6.36}$$

Where D is the elastic tensor, ψ and ϕ are the flow direction and the yield direction, and h is the plastic modulus. The stability condition of this constitutive relation can be evaluated by introducing the normalized second order work [15]:

$$d^2W \equiv \frac{d\tilde{\sigma} \cdot d\tilde{\epsilon}}{|d\tilde{\sigma}|^2} \tag{6.37}$$

Then, the Hill condition of stability can be obtained by replacing Eq. (6.36) in this expression. This results in

$$d^2W = R(\theta) + \frac{\langle\cos(\theta + \phi)\rangle}{h}\cos(\theta + \psi) > 0 \tag{6.38}$$

where $R(\theta)$ is defined by Eq. (6.15). In the case where the Drucker normality postulate $\psi = \phi$ is valid, Eq. (6.38) is strictly positive and, therefore, the Hill stability condition would be valid for all the stress states inside the plastic limit surface. On the contrary, for a non-associated flow rule as in our model, the second order work is not strictly positive for all the load directions, and it can take zero values inside the plastic limit surface (i.e. during the hardening regime where $h > 0$).

To analyze these instability, we adopt a circular representation of d^2W shown in Fig. 6.18. The dashed circles in these figures represent those regions where $d^2W < 0$. For stress ratios below $q/p = 0.7$ we found that the second order work is strictly positive. Thus, according to the Hill stability condition, this region corresponds to stable states. For the stress ratio $q/p = 0.8$, the second order work becomes negative between $27° < \theta < 36°$ and $206° < \theta < 225°$. This leads to a domain of the stress space strictly inside the plastic limit surface where the Hill condition of stability is not fulfilled, and therefore the material is potentially unstable.

As presented in Chapter 4, numerical simulations of biaxial tests show that strain localization is the most typical mode of failure. The fact that it appears before the sample reaches the plastic limit surface suggests that the appearance of this instability is not completely determined by the current macroscopic stress of the material, as it is predicted by the Drucker-Prager

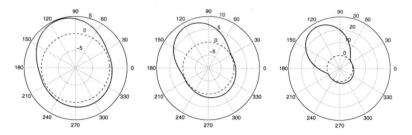

Figure 6.18: The solid lines show the second order work as a function of the direction of load for three different stress ratios $q/p = 0.6$ (left), 0.7 (center), and 0.8 (right) with pressure $p = 160kPa$. The dashed circles enclose the region where $d^2W < 0$.

theory. More recent analytic [81] and experimental [12, 53] works have focused on the role of the microstructure on the localized instabilities. Frictional slips at the particles have been used to define additional degrees of freedom [81]. The introduction of the particle diameter in the constitutive relations results in a correct prediction of the shear band thickness. The new degrees of freedom of these constitutive models are still not clearly connected to the micromechanical variables of the grains, but with the development of numerical simulations this aspect can be better understood.

6.5 Concluding remarks

The elasto-plastic response of a Voronoi tessellated sample of polygons has been calculated in the case of monotonic and quasistatic loading. The plastic response reflects several aspects of realistic soils. They have been discussed in relation to the existing elasto-plastic models. The most salient features are shown in Fig.6.19:

- The incremental elastic response has a centered ellipse as an envelope response. Below the stress ratio $q/p < 0.4$, this response can be described by the two material parameters of Hooke's law of elasticity: the Young modulus and the Poisson ratio. Above this stress ratio there is a dependence of the stiffness on the stress ratio, which can be connected to the anisotropy induced in the contact network during

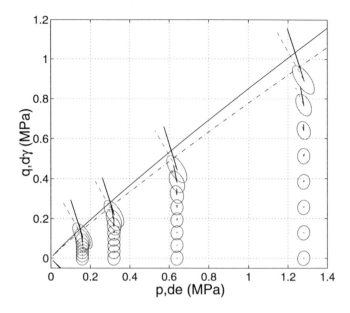

Figure 6.19: Elastic response $d\bar{\epsilon}^e$ and plastic response $d\bar{\epsilon}^p$ for different stress values. The yield direction is represented by the direction of the dashed line at each stress value. The solid line represents the plastic limit surface. The dash-dotted line is the failure surface.

loading. We should state that this result might be dependent on the preparation procedure. In particular, samples with void ratio different from zero show a smooth transition to the anisotropy, which requires further studies.

- The plastic envelope responses lie almost on the straight line defining the plastic flow direction ψ. The yield direction ψ and the plastic modulus h have also been calculated directly from the plastic response. In agreement with soil experiments, we found a non-associated flow rule of plasticity with $\psi < \phi$. This flow rule is in agreement with the prediction of the standard theory of elasto-plasticity.

- The flow direction and yield direction depend on the stress ratio, in agreement with the so-called stress-dilatancy relation of experiments on soils. In particular, the plastic flow for zero stress ratio has a nonzero deviatoric component suggesting an anisotropy induced for extremely small deviatoric strains. We found that this effect comes from the fact that the sliding contacts depart from the anisotropy when the sample is sheared.

- In the investigation of the connection between the plastic deformation and the number of sliding contacts, we found that the plastic modulus h decays exponentially as the fraction of sliding contacts increases. This contrasts with the linear decrease of the Young modulus E with the increase of the number of open contacts, suggesting that the deformation of the granular assembly is concentrated at the sliding contacts.

- The experimental method proposed by Tatsouka has been implemented to identify the yield surface. The resulting strain response shows that the transition from elasticity to elasto-plasticity is not as sharp as the Drucker-Prager theory predicts, but a smooth transition occurs. The fact that there is no purely elastic regime leads to the open question of how to determine the dependence of the response of soils on the history of the deformation.

- The calculation of the plastic limit condition and the failure surface shows that the failure appears during the hardening regime $h > 0$. This result is consistent with the Hill condition of stability, which states that for non-associated materials the instabilities can appear before the plastic limit surface.

Since the mechanical response of the granular sample is represented by a collective response of all the contacts, it is expected that the constitutive relation can be completely characterized by the inclusion of some internal variables, containing the information about the microstructural arrangements between the grains. We have introduced some internal variables measuring the anisotropy of the contact force network. The fabric coefficients a_i, measuring the anisotropy of the network of all the contacts, prove to be connected to the anisotropic stiffness. On the other hand, the fabric coefficients c_i, measuring the anisotropy of the sliding contacts, are related to the plasticity features of the granular materials.

Future work should be oriented towards the elaboration of a theoretical framework connecting the constitutive relation to these fabric coefficients.

To provide a complete micromechanically based description of the elasto-plastic features, the evolution equations of these internal variables must be included in this formalism. This theory would be an extension of the ideas which have been proposed to relate the fabric tensor to the constitutive relation of granular materials [3–6, 82].

Chapter 7

Granular Ratcheting

"... the micromechanical ratcheting, i.e. the systematic shift of contacts against each other due to geometrical asymmetry generated under the cyclic loading. This ratcheting can be macromechanically measurable by a slow convection movement within the packing."

Hans Herrmann: draft of the DFG project:
Micromechancial investigation of the granular ratcheting.

"If this effect is true, then it has a big implication for theories (and application) of constitutive laws in granular material. It is still difficult for me to believe it, and I will try to find time to do some tests of my own."

Peter Cundall: private communication.

" There have been a number of models which show very clear ratcheting in small cycles - with high stiffness on unloading and low stiffness on reloading. I think this is not what the experiments show. In reality one sees hysteretic response even for very small cycles - so there is energy dissipation - but there may actually not be much accumulation of strain. And one expects that if the density increases as a result of accumulation of volumetric compression, then that will tend to increase plastic and elastic stiffness and reduce the rate at which subsequent strains develop."

David Wood: private communication.

In this last chapter we will introduce a long time effect in granular materials, which is still under discussion in the scientific and engineering community. This effect is known as *ratcheting*, and it concerns the linear accumulation of permanent deformation per cycle in granular materials when they are subjected to load-unload stress cycles with extremely small loading amplitudes. Although there is wide experimental evidence about accumulation of permanent deformation under cyclic loading [14, 83–89], it is not clear whether this effect remains for small loading amplitudes, or if there is a certain regime where the material behaves perfectly elastic [28, 65, 68]. It is still also not clearly understood what is the role of the micromechanical rearrangements such as sliding, crushing and wearing of the grains, in the macromechanical aspects of the accumulation of plastic deformation with the number of cycles [83–86, 90, 91].

Here we will present numerical evidence of this ratcheting effect for small loading amplitudes on assemblies of densely packed polygons. This can be detected at the micromechanical level by a ratchetlike behavior at the contacts. This effect excludes the existence of the rather questionable finite elastic regime of noncohesive granular materials.

Before going to the results, we will introduce the concepts of *ratchet*, *ratchet effect* and *ratcheting*, which have been used in the recent years in many different contexts.

7.1 Ratchets and ratcheting

Chapter 46 of the Feynman Lectures on Physics [92] contains a celebrated illustration of a simple device which is able to extract work from unbiased thermal fluctuations. As shown in Fig. 7.1, the device is nothing but a pawl that engages the sloping teeth of a wheel, permitting motion in one direction only. An axle connects this wheel with some vanes, which are surrounded by a gas. The vanes are randomly hit by the gas molecules, but due to the presence of the pawl, only collision in one direction can make the wheel lift the pawl and advance it to the next notch.

The possibility to extract work from noise using ratchet devices has at-

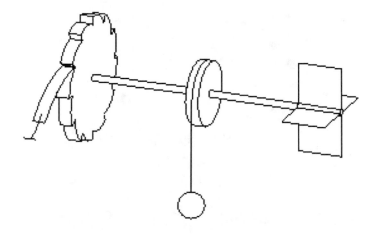

Figure 7.1: There are two boxes with a vane in one and a wheel that can only turn one way. Each box is in a thermal bath of gas molecules at equilibrium. The two boxes are connected mechanically by a thermally insulated axle. The entire device is considered to be reduced to microscopic size so gas molecules can randomly bombard the vane, to produce motion.

tracted many recent researchers [93, 94]. There is already an extensive body of work on this subject, driven by the need to understand the molecular motors that are responsible for many biological motions, such as cellular transport or muscle contraction [93]. Recently, this kind of mechanism has been experimentally demonstrated using the technology available to build micrometer scale structures. Many man-made ratchet devices have been constructed, and they are used as mechanical and electrical rectifiers [94].

Granular media also show ratcheting effects when they are vertically vibrated upon an asymmetric sawtooth-shaped base [95–97]. The main ingredients of the experimental setup are shown in Fig. 7.2. The base is vertically vibrated with a displacement that depends sinusoidally on time. Above a certain characteristic vibration amplitude, the asymmetry of the teeth breaks the symmetry of the AC driving force, leading to net horizontal motion [95]. Such rectification of a fluctuating force induces segrega-

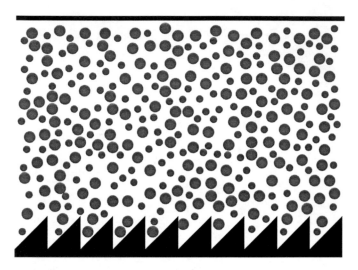

Figure 7.2: Diagram of the experimental apparatus to study the ratchet-induced flow in a granular material. The profile has a sawtooth shape. When the assembly is subjected to vertical vibration a convective flow in the horizontal direction appears.

tion and transport in the granular material [96, 97].

The concept of *ratcheting* has been also introduced in soil mechanics, to describe the gradual accumulation of permanent deformation in granular materials subjected to cyclic loading [83]. At the micromechanical level, it is related to a systematic shift of the sliding contacts. We will show that this is due to the load-unload asymmetry of the contact force network at each contact. This ratcheting can be macromechanically measurable by slow convection movement within the packing.

Before going into the micromechanical description of this effect, we will introduce the importance of the ratcheting in deterioration of structures and the existing theoretical approaches to this phenomenon.

7.2   Cyclic loading in soil mechanics

The ancient city of Petra was built from 800 BC to 100 AD by the Nabetean Arabs. In this era, Petra was a fortress, carved out of craggy rocks in an area, which was virtually inaccessible. In the first and second century, after the Romans took over, the city reached the peak of its fame. When caravan routes were slowly displaced by shipping, the city's importance gradually decreased; it fell into disuse and was lost to the world until 1812, when it was rediscovered by the Swiss explorer Johann Ludwig Burckhardt. Nowadays Petra is Jordan's number one tourist attraction. As a consequence, it is now in grave danger of being destroyed by the unstoppable march of tourism. More than 4000 tourists a day visit Petra's rocky tomb.

It is not just Petra's temples that are under threat of destruction. More than 600 millions tourists a year now travel the globe, and vast number of them wanting to visit the word's most treasures sites. If appropriate measures are not taken in time, tourism would certainly progressively destroy all these cultural treasures.

Not only tourism, but also transportation needs in general have taken off in the last decades [90]. Traditional methods to evaluate the deterioration of foundations under repeated loading are still almost completely empirical [86]. The increase of traffic loads have resulted in a rapid deterioration of the public road system, and therefore in a rise of maintenance expenses. This has attracted the attention of public authorities that is urging the road construction industry to optimize its designs.

Concerning the pavement design, experimental [14, 86–88, 98] and analytical [28, 90] procedures have been developed based on the analysis of the response of the unbound granular materials under specific load conditions. These studies are based on the assumption that rutting occurs mostly in the unbound granular materials used in the subgrades [98]. The gradual accumulation of permanent deformations in the subgrade can lead to eventual formation of ruts or cracks in the pavement due to excessive rutting. Whether a given system will experience progressive accumulation of permanent deformation, or whether the increase of permanent deformation will stop, is crucial for performance predictions.

Most of the research carried out over recent years concentrated on the overall behavior of prepared samples in laboratory tests which was then interpreted using constitutive models, maybe due to the practical difficulties in studying permanent deformation at the grain level. Recent pavement designs, however, point out the necessity of a detailed study on the micromechanics of the permanent strain.

7.2.1   Continuous models for cyclic loading

The constitutive behavior of unbound granular soil under cyclic loading has been investigated in the framework of the *shakedown theory* [28, 89, 90]. This theory predicts that a granular material is liable to show progressive accumulation of plastic strains under repeated loading if the magnitude of the applied loads exceeds a limiting value called the shakedown limit. The material is then said to exhibit *Ratcheting* On the other hand, if the loads are under this limit, the growth of permanent deformations will eventually level off and the material is said to have attained a state of shakedown by means of adaptation to the applied loads. More in detail, the shakedown concept maintains that there are four categories of material response under repeated loading:

- An *elastic* range for low enough loading levels, in which no permanent strains occurs.

- *Elastic shakedown*, where the applied stress is slightly under the plastic shakedown limit. The material response is plastic for a finite number cycles. However the ultimate response is elastic.

- *Plastic shakedown*, where the applied stress is slightly less than that required to produce ratcheting. The material achieves a long-term steady state response with no accumulation of plastic strain and hysteresis.

- *Incremental collapse or ratcheting*, where the applied repeated stress is relatively large. Plastic strains accumulate rapidly with failure occurring in the relatively short term.

Shakedown theory is essentially an extension of the classical Drucker-Prager theory of elasto-plasticity. This theory describes the cyclic loading response by postulating a certain region in the stress space where only elastic deformations are possible [64]. However, this basic assumption does not seem to be confirmed by experiments on cyclic loading, which show that the onset of the ratcheting with the increase of the loading amplitude is gradual and not sharply defined [14].

Some sophisticated models have been proposed in order to mend these deficiencies of the Drucker-Prager theory in the description of the smooth transition from elasticity to elasto-plasticity. In the boundary surface theory, the cumulative plastic deformation for small cycles of loading is modeled by shrinking the elastic nucleus to the current stress state [77]. This theory is not found widespread in the geotechnical application, due to its complex mathematical structure which does not allow one to simulate large number of cycles, and the great number of parameters in it that are difficult to calibrate.

Taking another perspective, some cyclic loading models have been developed starting from the theory of hypoplasticity. Besides the stress and the void ratio, these models introduce additional internal variables such as the back stress tensor [99] or the intergranular strain [100]. These models have also been skeptically received by the engineering community due to the scarce physical meaning of these internal variables.

Most of the attempts to identify the internal variables of constitutive equations are based on macromechanical observations of the response of soil samples in conventional apparatus. The micromechanical investigation would certainly help get an insight into these internal variables. Indeed, the mechanical response of the granular soils is no more than a combined response of many micromechanical arrangements, such as interparticle slips, breakage of grains and wearing of the contacts.

7.2.2   Discrete approach on the cyclic loading response

Using discrete element models, different micromechanic aspects of the response of granular materials under monotonic loading have been adressed by many authors. Amazingly, few studies have been reported about the behavior of granular material under cyclic loading conditions. Some recent discrete element calculations of cyclic loading have addressed the phenomenon of liquefaction [101]. This is a phenomenon that takes place during earthquakes. It causes a reduction of the stiffness of soils so that they behave as viscous fluids rather than solids. These numerical simulations are performed under strain-controlled loading. Under small loading amplitudes, there is a certain amount of stress buildup over the course of the loading cycles, but this effect should stop after some number of cycles because the contact forces cannot increase indefinitely.

In order to observe ratcheting behavior, unless one of the directions of the sample must be subjected to stress controlled loading. In this way, permanent deformation is allowed in this direction. We perform here a simulation of load-unload stress cycles. This condition is similar to experimental tests performed for testing granular materials for pavement [14, 90].

7.3   Simulation of cyclic loading

Just to start a micromechanical investigation on the behavior of soils under cyclic loading, we perform MD simulations on polygonal packings. To obtain homogeneous, dense granular samples, the polygons are placed randomly inside a rectangular frame consisting of four walls. Then, a gravitational field is applied and the sample is allowed to consolidate. The external load is imposed by applying a force $\sigma_1 H$ and $\sigma_2 W$ on the horizontal and vertical walls, respectively. Here σ_1 and σ_2 are the vertical and horizontal stresses. H and W are the height and the width of the sample.

In the simulation of the cyclic loading response of a polygonal packing we use a procedure equivalent to the laboratory biaxial test. First, the sample is isotropically compressed until the pressure p_0 is reached. Then, the vertical stress $\sigma_1 = p_0$ is kept constant and the horizontal stress is modulated as $\sigma_1 = p_0 + \Delta\sigma[1 - \cos(\pi t/t_0)]/2$. This smooth modulation is chosen in order to minimize the acoustic waves produced during the load-unload transition. $\Delta\sigma$ is chosen between $0.001p_0$ and $0.6p_0$. These values should be compared to the maximal stress during the biaxial test, which is around $0.75p_0$ in this model.

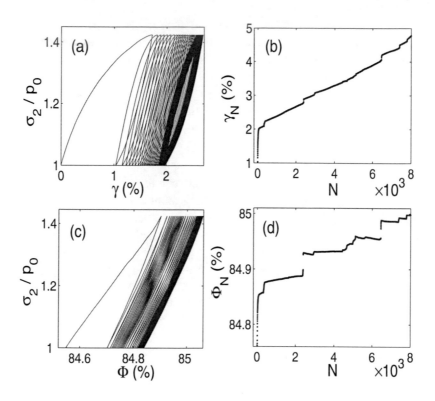

Figure 7.3: (a) Deviatoric stress versus deviatoric strain in the first 40 cycles. (b) permanent (plastic) strain γ_N after N cycles versus the number of cycles. (c) stress against the volume fraction in the first 40 cycles. (d) volume fraction Φ_N after N cycles versus number of cycles.

7.3.1   Stress-strain calculation.

In experimental tests, the response of a sample subjected to loading-unloading stress cycles is given by a progressive compaction, and a permanent accumulation of plastic deviatoric deformation as the number of cycles increases. We will see that these two important features are observed in our numerical simulations.

The strain tensor is calculated here averaged over a representative volume element (RVE). This RVE is obtained selecting the polygons whose centers of mass are less than 10ℓ from the center of the sample, where ℓ is the mean diameter of the polygons. Then, the strain is calculated as the displacement gradient tensor averaged over the area enclosed by the initial configuration of these polygons. From the eigenvalues ϵ_1 and ϵ_2 of the symmetric part of this tensor (usually called strain tensor) we obtain the deviatoric strain as $\gamma = \epsilon_1 - \epsilon_2$. The volume fraction is calculated as $\Phi = (V_p - V_0)/V_b$, where V_p is the sum of the areas of the polygons, V_0 the sum of the overlapping areas between them, and V_b the area of the rectangular box.

Part (a) of Fig. 7.3 shows the relation between the axial stress σ_1 and the deviatoric strain γ in the case of a loading amplitude $\Delta\sigma = 0.6p_0$ where $p_0 = 160kPa$. This relation consists of open hysteresis loops, which narrow as consecutive load-unload cycles are applied. This hysteresis produces an accumulation of strain with the number of cycles which is represented by γ_N in part (b) of Fig 7.3. We observe that the strain response consists of short time regimes, with rapid accumulation of plastic strain, and long time *ratcheting* regimes, with a constant accumulation rate of plastic strain of around 2.4×10^{-6} per cycle.

Part (c) of Fig. 7.3 shows the relation between the deviatoric stress and the volume fraction. This consists of asymmetric compaction-dilation cycles, which make the sample compact during the cyclic loading. This compaction is shown in part (d) of Fig. 7.3. We observe a slow variation of the volume fraction during the *ratcheting* regime, and a rapid compaction during the transition between two ratcheting regimes. Note that the amount of ratcheting, i.e. the slope of the curve in part (b) of Fig. 7.3, shows no dependence with the compaction level of the sample. This suggests that the granular ratcheting will remain for very large number of cycles, even when the volume ratio is very close to the saturation level.

The evolution of the volume ratio seems to be rather sensitive to the initial random structure of the polygons. Even so we found that after 8×10^3 cycles the volume fraction still slowly increases in all the samples. This behavior resembles the very slow compaction that has been experimentally observed during the cyclic shearing on packing of spheres [102]. In these experiments, the convergence of the volume ratio to the saturated level proves to be slower than any exponential or algebraic law.

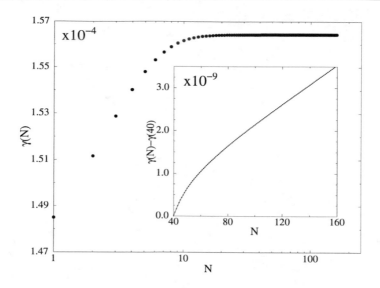

Figure 7.4: Cumulative plastic deformation as a function of the number or cycles for $\Delta\sigma = 0.01p_0$. The magnification shows the strain accumulated after the $40th$ cycle.

The extremely slow dynamics in the evolution of the granular packing shows an astonishing analogy with the behavior of glassy systems [103]. Based on a considerable amount of experimental data of compaction of granular materials, the similarity in the dynamics of granular matter under vibration and glass forming materials has been addressed by several authors [104–106]. This has been first revealed by the very slow relaxation of the density. Later on, memory experiments [102] and simulations inspired by earlier spin glass studies [107, 108] have also given support to this conclusion.

7.3.2 Limit of small cycles

One would expect that for small enough amplitudes of the loading cycles, one can reach the elastic regime postulated in the shakedown theory [28]. In an attempt to detect this elastic regime, we decreased the amplitude of the load cycles and evaluated the corresponding asymptotic response.

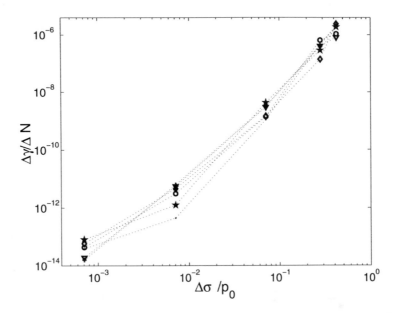

Figure 7.5: Plastic deformation per cycle for different loading amplitudes. The calculations are performed on six different samples.

Fig.7.4 shows the cumulative plastic deviatoric strain γ resulting from the application of loading cycles with amplitude $\Delta\sigma = 0.01p_0$. During the first cycles a transient regime showing a decay of the permanent deformation per cycle is observed. This behavior resembles the *shakedown* response of the elasto-plastic models. However, a magnification of Fig.7.4 reveals a surprising fact: After the application of hundred cycles, the shakedown behavior is replaced by the ratcheting regime. In this asymptotic behavior, one obtains a constant amount of plastic deformation in each cycle.

Regardless of the amplitude of the loading cycles, one always obtains ratcheting behavior in the long time behavior. This is shown in the accumulation strain rate $\Delta\gamma/\Delta N$ for different loading amplitudes $\Delta\sigma$ in Fig. 7.5. A constant accumulation of strain is observed during the cyclic loading, even when the amplitude is as small as 10^{-3} times the applied pressure. Of course, due the smallness of the ratcheting response for these loading amplitudes, one can say that for small loading amplitudes the response is practically elastic. Even if the slight repeated loading produced by the

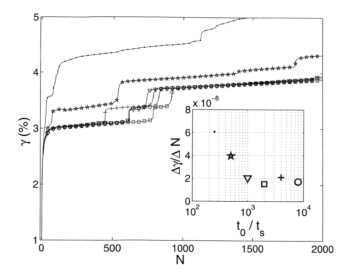

Figure 7.6: Permanent deviatoric strain for different periods of cyclic loading t_0. The inset shows the plastic deformation per cycle averaged over the last 1000 cycles. Each symbol in the inset corresponds to a value of t_0/t_s, where t_s is defined in Sec. 3.6.

transit of ants would produce plastic deformation after some centuries, it is not possible to make them to follow the same path all this time. However, it is important to address that Fig. 7.5 shows a smooth transition from the shakedown response to the ratcheting response. In the context of the phase transitions, this means that the distinction between the ratcheting and shakedown regime is rather meaningless.

7.3.3 Quasi-static limit

Since the molecular dynamics involves damping forces, it is important to know what is the role of these forces in the granular ratcheting behavior. Damping and inertial effects can be evaluated by performing the same test with different loading frequencies. Fig. 7.6 shows that as the frequency is reduced, the ratcheting effect gets progressively smaller until the quasistatic regime is reached. In this regime a reduction by one half of the frequency does not affect the strain response more than 5%. From this

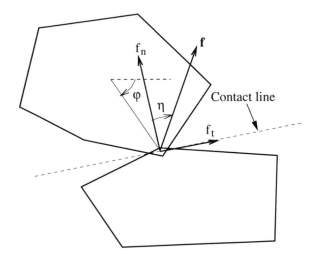

Figure 7.7: Orientation angle φ and mobilized angle η of the contact force **f**.

result one can conclude that damping or inertial effects do not affect the appearance of ratcheting in the sample, so that this is a genuine quasistatic effect.

Note that the time in which the transition between two ratcheting regimes occurs, seems to be different according to the frequency. Thus, damping or inertial effect may be important to include in the description of this transition. This study is however beyond to the scope of this work.

7.4   Micromechanical aspects

Due to the strong temporal fluctuations that have been observed in driven granular materials [109], the existence of these ratcheting regimes with constant accumulation of plastic deformation per cycle appears to be somewhat counterintuitive. We have noticed, however, that the existence of quasiperiodic regimes in the evolution of the contact forces can explain this particular behavior.

The basic elements of the micromechanical description of the granular ratcheting are shown in Fig. 7.7. For each contact we define an angle

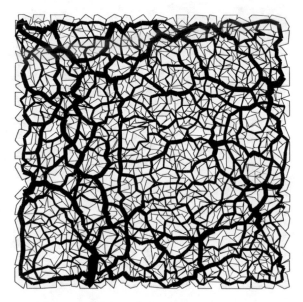

Figure 7.8: Contact force network in an isotropically compressed sample. The width of the lines represents the normal force.

φ, that is given by the orientation of the branch vector. This vector connects the center of mass of the polygon with the point of application of the force. The contact force $\mathbf{f}$ is decomposed in its normal f_n and tangential f_t components respect to the contact line. The angle $\eta = \arctan(f_t/f_n)$ is defined as the mobilized angle of the force. The sliding condition is given by $\tan(\eta) = \pm\mu$, where μ is the friction coefficient.

7.4.1 Fluctuations on the force.

A striking feature of granular materials is that distribution of forces within the material shows to be very heterogeneous. As shown in Fig. 7.8, the stress applied on the boundary is transmitted through chains along which the contact forces are particularly strong. These heterogeneities have also been observed using numerical simulations [8], and experimentally, using photo-elastic experiments [74, 75].

We have first studied the evolution of the distribution of the normal forces

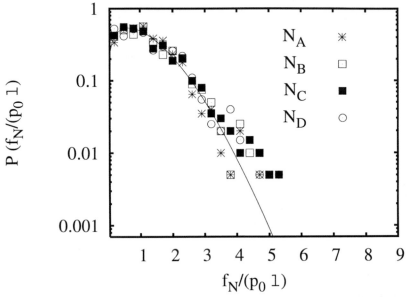

Figure 7.9: Distribution function of the normal forces in the contacts, measured at four different times of the simulation: $N_A = 10$, $N_B = 105$, $N_C = 150$ and $N_D = 190$. In this simulation $\Delta\sigma = 0.1$. The equation of the best fit curve is: $0.843x^{0.431} \exp{-0.546x^{1.630}}$.

during the cyclic loading, in the case of a loading amplitude $\Delta\sigma = 0.6p_0$. A broadening of the distribution is observed during each loading phase, followed by a narrowing of the distribution during the unloading phase. The time evolution of the first and the second moment of the distribution show that it reaches a periodic broadening-narrowing regime once the ratcheting behavior is reached.

In Figure 7.9 we plot the distribution function of normal forces at four different snapshots of the simulation. The best-fit curve is also included for an easier comparison. Note that although all distributions were measured at different times of the simulation, they correspond to the same stage of the cyclic loading. It is observed that the shape of the distribution of forces at this point remains approximately constant throughout the whole simulation. In this work we do not study the evolution of this distribution in detail, but rather focus on the evolution of the sliding contacts.

7.4.2   Sliding contacts

One of the most important features of the force network is the high number of sliding contacts. Although most of the contacts satisfy the elastic condition $|f_t| < \mu f_n$, the strong heterogeneities of the contact force network produce a considerable amount of contacts reaching the sliding condition $|f_t| = \mu f_n$ during the compression. Those sliding contacts carry most of the irreversible deformation of the granular assembly during the cyclic loading. Opening and closure of contacts are quite rare events, and the coordination number of the packing keeps it approximately its initial value 4.43 ± 0.08 in all the simulations.

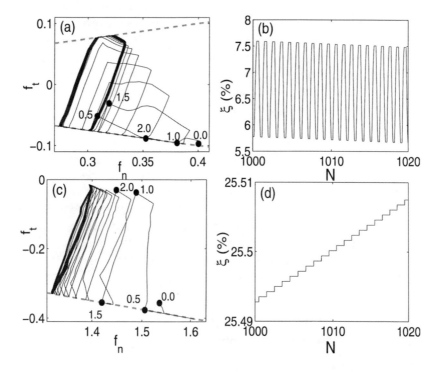

Figure 7.10: (a) and (b) Trajectories of the contact force of two selected sliding contacts. The dots denote the times $t = 0, 0.5t_0, ..., 2t_0$ in unit of the period t_0. The dashed line shows the sliding condition $|f_t| = \mu f_n$. (b) and (d) Plastic deformation ξ at the contacts shown in (a) and (c).

Some typical trajectories of the force at the sliding contacts are shown in parts (a) and (c) of Fig. 7.10. After certain loading cycles the contact forces reach the quasi-periodic behavior. In this regime, a fraction of the contacts reaches almost periodically the sliding condition. The load-unload asymmetry of the contact force loops makes the contacts slip the same amount and in the same direction during each loading cycle.

A measure for the plastic deformation of the sliding contact is given by $\xi = (\Delta x_t^c - \Delta x_t^e)/\ell$, where Δx_t^c and Δx_t^e are the total and the elastic part of the tangential displacement at the contact, the last one being given by Eq. (3.1) in Sec. 3.2.2. Parts (b) and (d) of Fig. 7.10 show the plastic deformation ξ of the two sliding contacts. Due to the load-unload asymmetry of the contact force loop, a net accumulation of plastic deformation is observed in each cycle. In the case of the contact shown in part (b) of Fig. 7.10, the contact slips forward during the loading, and backward during the unloading phase. This sliding results in a net accumulation of permanent deformation per cycle. The other contact behaves elastic during the loading and slips during the unloading. This mechanism resembles the Feynman ratchets presented in Sec. 7.1.

It is interesting to observe the spatial correlation of these ratchets. Fig. 7.11 shows a snapshot of the field of plastic displacement per cycle at the contacts inside of the assembly. We see that correlated displacements coexist with a strongly nonhomogeneous distribution of amplitudes. Localized slip zones appear periodically during each ratcheting regime. Some slip zones are destroyed and new ones are created during the transition between two ratcheting regimes. Moreover, we notice that these ratchets are found as well at the boundaries as in bulk material, without the layering effects observed in vibrated granular materials [75].

7.4.3   Anisotropy & Feynman ratchets

From small loading amplitudes, the appearance of ratchetlike motion appears to be a consequence of the anisotropy induced by the loading on the distribution of the sliding contacts. We will perform a micromechanical inspection of this effect in the case of $\Delta\sigma = 0.01p_0$.

The anisotropy of the sliding contact can be measured from the orientational distribution of these contacts. This distribution is given by the orientation φ of the branch vectors of the sliding contacts (see Fig. 7.7). During compression, the distribution of sliding contacts is isotropic. However, we found in Sec. 6.8 that extremely small loads induce anisotropy. Indeed, during loading those sliding contacts whose orientation is nearly parallel to the loading direction leave the sliding condition.

The appearance of the anisotropy can be schematically explained from Fig. 7.7. Let us suppose that both polygons belong to an assembly, which has been isotropically compressed. Let us also assume that the contact force satisfies the sliding condition $f_t = \mu f_n$. Imagine that a small loading is imposed on the assembly in the vertical direction. Since the branch vector

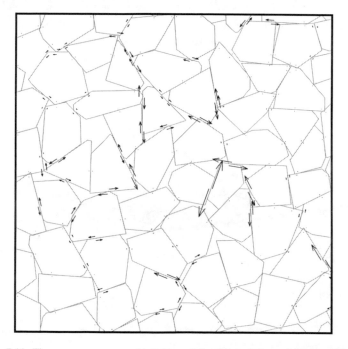

Figure 7.11: The arrows represents the field $\mathbf{u}$ of the plastic deformations accumulated at the contacts during one cycle: $\mathbf{u} = 500(\xi_{N+1} - \xi_N)$, where ξ_N is the plastic displacement after N cycles.

in this example is oriented nearly in this loading direction, the normal force will increase more than the tangential one, and the contact will leave the sliding condition. On the contrary, if the loading is applied in the horizontal direction, the tangential force will increase more than the normal force, and the contact will remain in the sliding condition.

This picture is useful to explain the complex evolution of the orientational distribution of the sliding contacts, that is shown in Fig. 7.12. During the first cycle, sliding contacts oriented nearly parallel to the load direction stick during the loading phase, and some of them slip during the unload phase. On the other hand, the sliding contacts orientated nearly perpendicular to the load direction slip during the loading phase, and stick during the unload phase. These slip-stick mechanisms in each load-unload cycle resemble again the Feynman ratchets. We will see that after many loadings some of the initially sliding contacts still reach the sliding condition, even under extremely small loading amplitudes. The ratchetlike behavior of these contacts gives rise to a constant accumulation of permanent deformation per cycle in the material.

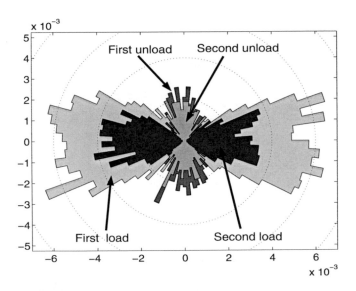

Figure 7.12: Distribution of the orientation φ of the sliding contacts arising in the first two load-unload phases.

7.4.4 Displacement field

During the ratcheting regime, the constant accumulation of plastic deformation per cycle at the sliding contacts will be reflected in a constant displacement per cycle at the individual grains. It is of great interest to study the patterns that are created by the displacement field of all the grains.

During the cyclic loading, the trajectory of a single particle is given by a constant, small displacement per cycle in the ratcheting regime, and a large displacement during the transition between two ratcheting regimes. Typically, the maximal displacement per cycle at this transition is one or two orders of magnitude larger than in the ratcheting regime.

The upper part of Fig. 7.13 shows a snapshot of the displacement per cycle of the particles for these two cases. The most important remark of this flow is the formation of vortex structures. An animation of this flow showes a constant vorticity field during the ratcheting regime, and large vorticities during the transition of two ratcheting regime. We have also observed that vortex structures are created and destroyed during this transition.

Since the vorticity is linked with the a nonvanishing antisymmetric part of the displacement gradient [110], the strain tensor is not sufficient to provide a complete description of this convective motion during cyclic loading. An appropriate continuum description would require the introduction of additional degrees of freedom taking into account the vorticity. As in the case of the shear band formation, the Cosserat theory may be a good alternative [111].

7.4.5 Micro-macro transition

In this last chapter we established a correlation between the amount of the plastic deformation and the fraction of sliding contacts. Coming back to this point, we will see that the main aspects of the hysteretic response during cyclic loading can be explained from the analysis of the sliding contacts.

We will establish a correlation between the dynamics of the sliding con-

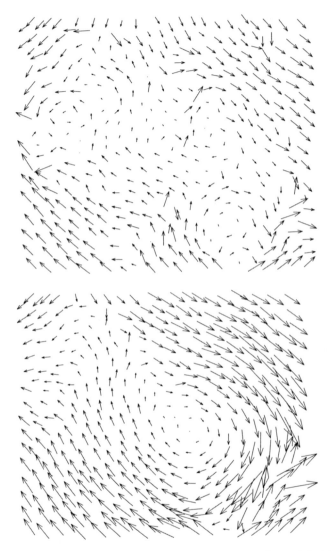

Figure 7.13: Snapshot of the simulation of cyclic loading with $\Delta\sigma = 0.6p_0$.The upper image corresponds to an instant in which the assembly is in a ratcheting regime; the lower one to a instant during the transition between two ratcheting regimes. The arrows represent $10^5\Delta u$ in the upper image and $10^3\Delta u$ in the lower one. Here Δu is the displacement of the particle per cycle.

tacts and the evolution of the stiffness of the material. The latter is given by the slope of the stress strain curve in part (a) of Fig. 7.3. The evolution of the fraction $n_s = N_s/N_c$ of sliding contacts with the number of loading cycles is shown in Fig. 7.14. Here N_s is the number of sliding contacts and N_c is the total number of contacts. During each loading phase, the number of sliding contacts increases, giving rise to a continuous decrease of the stiffness as shown in part (a) of Fig. 7.3.

A very important aspect of the dynamic of the sliding contacts is the abrupt reduction in the number of sliding contacts at the transition from load to unload. At the macromechanical level, this is reflected by the typical discontinuity in the stiffness observed under reversal loading.

During cyclic loading the number of sliding contacts tends to decrease, which produces a narrowing of the hysteresis loops. In the long time behavior one can also see that some contacts reach almost periodically the sliding state even for extremely small loading cycles. The ratchetlike behavior of these contacts produces a net displacement of the hysteretic stress-strain loop in each cycle, giving rise to the ratcheting response. Certainly, a deeper investigation of the evolution of this sliding contacts during loading would provide the basis for a micromechanical description of the hysteretic response of soils.

7.5 Concluding remarks

A grain scale investigation of the cyclic loading response of a packing of polygons has been presented. In the quasistatic regime, we have shown the existence of long time regimes with a constant accumulation of plastic deformation per cycle, due to ratcheting motion at the sliding contacts.

As the loading amplitude decreases, we observe a smooth transition from the ratcheting to the shakedown regimes, which does not allow one to identify a purely elastic regime. For small loading amplitudes the granular ratcheting results from the anisotropy induced by the loading on the sliding contacts.

The overall response of the polygonal packing under cyclic loading con-

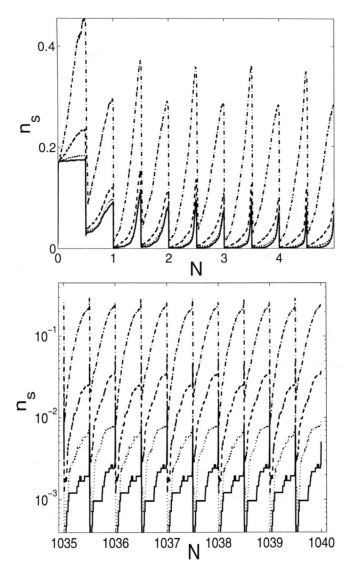

Figure 7.14: Fraction of sliding contacts n_s in the long time behavior for different values of $\Delta\sigma_y/p_0$: 0.424 (dash-dotted line), 0.0707 (dashed line), 0.00707 (dotted line) and 0.000707 (solid line)

sists of a sequence of long time ratcheting regimes, with slow accumulation of plastic deformation. These regimes are separated by short time regimes with large plastic deformations.

The analysis of the displacement field per cycle of the particles shows that each one moves with constant displacement per cycle during the ratcheting regimes. These displacements form vortexlike structures, which remain during the time of the ratcheting regime.

The existence of granular ratcheting may have deep implications in the study of the permanent deformation of soils subjected to cyclic loading. More precisely, it may be necessary to introduce internal variables in the constitutive relations, connecting the dynamics of the sliding contacts to the evolution of the continuous variables during cyclic loading.

At this time, a comparison of the dynamic simulations with realistic situations is limited by the computer time needed for simulations. Using a computer with a $2.4GHz$ processor we are able to simulate only 20 cycles per hour. The improvement of computational efficiency may require one to explore another discrete element techniques such as the method of contact dynamic [9]. Contact dynamics would be a more appropriate method for the simulation of these systems, especially in the case of grains with very high stiffness.

The similarity of results with the recently reported elasto-plastic behavior in packings of disks [112] indicates that these phenomena do not depend on the geometry of the grains, and that they may be inherent to the granular interactions. the existence of granular ratcheting in three-dimensional systems is still an open question.

Chapter 8

Conclusions

In this thesis a micromechanical investigation of the plastic deformation of soils has been presented, using molecular dynamics simulations. A simple two-dimensional model has been used to represent the granular material. This model captures the diversity of grain shapes, as well as the quasistatic friction forces at the contacts. An averaging formalism has been implemented in order to compute the macromechanical quantities such as the stress and strain tensor, from the micromechanical quantities of the simulations: contact forces, displacements and rotations of the grains.

The incremental stress-strain relation of this model has been calculated in the quasistatic regime. The simulation results have been compared to the existing incremental rate-independent constitutive models. The resulting incremental response has been used to verify the basic assumption of the elasto-plastic theory and incremental nonlinear models. In spite of the simplicity of our model, it can reproduce the principal features of realistic soils, such as the anisotropy of the elastic response, the stress-dilatancy relation, the non-associated flow rule of plasticity, the strain localization, and the existence of instabilities in the hardening regime.

As elasto-plastic theories predict, the resulting incremental response has two well-defined tensorial zones. We found also that the superposition principle is fulfilled, which is consistent with the existence of these tensorial zones. These results suggest that the elasto-plasticity is more appropriate than the incremental nonlinear models, in the description of the incremental response of this model.

The connection between the elasto-plastic response and the micromechanical rearrangements has been studied by the introduction of some internal variables, measuring the anisotropy of the contact network. These variables correspond to a generalization of the fabric tensor. They have been used to correlate the anisotropy of the contact network with the elastic response. The most salient aspects of the plastic deformation are also explained in terms of the anisotropy induced by loading in the subnetwork of the sliding contacts.

Finally, we present a micromechanical investigation of the hysteretic response when the granular samples are subjected to load-unload stress cycles. We report on the existence of ratcheting regimes with a constant accumulation of permanent deformation per cycle. At the grain level, we have observed that some contacts reach almost periodically the sliding condition even for extremely small loading amplitudes. The ratchetlike behavior of these contacts produces a net displacement per cycle of the hysteretic stress-strain loop leading to an overall ratcheting response.

The most salient aspect of this ratcheting behavior is that it excludes the existence of a purely elastic regime. In fact, we found that as the loading amplitude decreases, the transition from the ratcheting to the shakedown response is rather smooth, which does not allow us to distinguish an elastic regime. A micromechanic inspection of the cyclic loading response has shown that any load involves sliding contacts, and hence, plastic deformation. Experimental studies on dry sand seem to show that the truly elastic region is probably extremely small. The elastic region that is used in Drucker-Prager theories in the modeling of soils seems to be a pragmatic compromise which helps to give a dependence of response on recent history, but is not a necessary feature.

In summary, two important conclusions can be drawn from the analysis of the quasistatic mechanical response of the polygonal samples:

- **The calculation of the incremental stress-strain relation leads to two well defined tensorial zones.**

- **It is not possible to define a finite region in the stress space where only elastic deformations are possible.**

These two conclusions appear to contradict both the Drucker-Prager theory and the hypoplastic models. In future work, it would be important to revisit the question of the incremental nonlinearity of soils from micromechanical considerations.

8.1   Outlook

In 1986 Dafalias introduced the concept of *hypoplasticity* [77]. This development was motivated by the necessity to describe the hysteretic response of soils under cyclic loading. Dafalias has shown that shrinking the elastic regime to the current stress point, one can reproduce the observed continuous transition from the elastic to the elasto-plastic behavior. This limit leads to a constitutive relation in terms of the bounding surface and some internal variables, which are the macromechanical manifestation of the material microstructure.

Following a different approach, Kolymbas [16] and Chambon [70] introduced a new concept of hypoplasticity, based on a historic independent, nonlinear incremental relations. Subsequent improvements have introduced certain tensorial quantities, which take into account the dependence of the mechanic response with the history of the deformation [99, 100].

Despite that these formulations are completely different, they seem to converge at the same point: the necessity to introduce internal variables to describe the essential feature of mechanics of granular materials, that any loading involves plastic deformation.

Most of the attempts to identify the internal variables of the constitutive relation have been based on observations of the response of soil samples in conventional tests [1]. The recent improvements in discrete element modeling (DEM) allow one to perform this investigation from the micromechanical point of view. We are in condition to develop a micromechanical models giving the internal variables of the constitutive models in terms of the microstructural information, such as polydispersity of the grains, fabric coefficients, and force distributions.

To start the micromechanical investigation of those internal variables, it

would be necessary to introduce an explicit relation between the incremental stress-strain relation and some statistics measuring the anisotropy
of the contact network and the fluctuations of the contact forces. One way
to do that is to introduce the statistic distribution $\Omega(\ell, \varphi, \mathbf{f})$ of the micromechanical variables. Here ℓ and φ are the magnitude and the orientation of
the vector connecting the center of mass of the grain with the point of application of the contact force $\mathbf{f}$. In the most general case, the incremental
stress-strain relation can be given by

$$d\sigma_{ij} = \int_{\boldsymbol{\lambda}} d\boldsymbol{\lambda} \Omega(\boldsymbol{\lambda}) R_{ijkl}(\boldsymbol{\lambda}) d\epsilon_{kl}. \tag{8.1}$$

Here $\boldsymbol{\lambda} = (\ell, \varphi, f_n, f_t)$ and R_{ijkl} is a tensorial quantity, taking into account
the local fluctuations of the deformation at the contacts with respect to the
principal value of the averaged incremental strain tensor $d\epsilon$ [113]. Note
that the marginal distribution of Ω contains the basic statistics which have
been intensively investigated in the microstructure of granular material:
the size distribution $\Omega(\ell)$ [35, 53, 114], anisotropy of the contact network
$\Omega(\varphi)$ [3–6] and the contact force distribution $\Omega(\mathbf{f})$ [7, 8, 75, 115]. A great
challenge is to find explicit expressions for the incremental stress-strain response in terms of internal variables, given as a function of this distribution
Ω. This investigation would be an extension of the ideas which have been
proposed to relate the fabric tensor to the constitutive relation [3–6, 82].

The traditional fabric tensor, measuring the distribution of the orientation
of the contacts, cannot fulfill a complete micromechanical description, because it does not make a distinction between elastic and sliding contacts
[4]. New structure tensors, taking into account the statistics of the subnetwork of the sliding contacts, must be introduced to give a micromechanical
basis to the plastic deformation. The identification of these internal variables, the determination of their evolution equations, and their connection
with the macroscopic variables would be a key step in the development of
an appropriate continuous description of granular materials.

The evolution equation for these internal variables could be determined
from the evolution equation of Ω during loading. This can be obtained
from the conservation equations of the contacts [116]:

$$\frac{\partial \Omega}{\partial t} + \frac{\partial (\Omega v_i)}{\partial \lambda_i} = Q(\lambda). \tag{8.2}$$

The velocity field $\mathbf{v}(\boldsymbol{\lambda}) = d\boldsymbol{\lambda}/dt$ can be investigated from DEM by following the evolution of the contacts during the simulation. The source term Q takes into account the contacts arising or disappearing during the deformation of the granular assembly, as a consequence of the rearrangement of the granular skeleton and the eventual fragmentation of the grains. In future work, an important goal would be to determine the role of such micromechanical rearrangements in the overall mechanical response of granular materials.

Let's conclude remarking that the statistical mechanics has been one of the most fundamental and successful theories of the matter. It allows one to explain many thermodynamic aspects of solids, liquids and gases from microscopic physical laws. Contrary to this, different statistical mechanical approaches intending to provide a micromechanical basis to the complex mechanical response of granular material have given few satisfactory results. In the author's opinion, a real advance in this field could be made by contending that granular materials belong to a new class of materials, which require their own theoretical framework. We attempted here to delineate a rigorous framework in order to derivate the incremental response of soils from strictly micromechanical considerations.

Bibliography

[1] G. Gudehus, F. Darve, and I. Vardoulakis. *Constitutive Relations of soils*. Balkema, Rotterdam, 1984.

[2] P. A. Cundall, A. Drescher, and O. D. L. Strack. Numerical experiments on granular assemblies; measurements and observations. In *IUTAM Conference on Deformation and Failure of Granular Materials*, pages 355–370, Delft, 1982. Balkema,Rotterdam.

[3] Y. Tobita. Importance of incremental nonlinearity in the deformation of granular materials. In *IUTAM Symposium on Mechanics of Granular and Porous Materials*, pages 139–150. Kluwer Academic Publishers, 1997.

[4] R. J. Bathurst and L. Rothenburg. Micromechanical aspects of isotropic granular assemblies with linear contact interactions. *J. Appl. Mech.*, 55:17–23, 1988.

[5] C. Thornton and D. J. Barnes. Computer simulated deformation of compact granular assemblies. *Acta Mechanica*, 64:45–61, 1986.

[6] M. Lätzel. *From discontinuous models towards a continuum description of granular media*. PhD thesis, Universität Stuttgart, 2002.

[7] K. Bagi. Statistical analysis of contact force components in random granular assemblies. *Granular Matter*, 5:45–54, 2003.

[8] F. Radjai, M. Jean, J. J. Moreau, and S. Roux. Force distribution in dense two-dimensional granular systems. *Phys. Rev. Lett.*, 77(2):274, 1996.

[9] J. J. Moreau. Some numerical methods in multibody dynamics: application to granular materials. *Eur. J. Mech. A*, 13:93–114, 1994.

[10] P. W. Rowe. The stress dilatancy relations for static equilibrium of an assembly of particles in contact. *Proc. Roy. Soc.*, A269:500–527, 1962.

[11] P. A. Vermeer. Non-associated plasticity for soils, concrete and rock. In *Physics of dry granular media - NATO ASI Series E350*, page 163, Dordrecht, 1998. Kluwer Academic Publishers.

[12] J. Desrues. *Localisation de la deformation plastique dans les materieux granulaires*. PhD thesis, University of Grenoble, 1984.

[13] P. A. Vermeer. The orientation of the shear bands in biaxial tests. *Geotechnique*, 40(2):223–236, 1990.

[14] S. Werkmeister, R. Numrich, A. R. Dawson, and F. Wellner. Deformation behaviour of granular material under repeated dynamic loading. In *Environmental Geomechanics*, Monte Verita, 2002. Presses Polytechniques et Universitaires Romandes.

[15] F. Darve and F. Laouafa. Instabilities in granular materials and application to landslides. *Mechanics of Cohesive-Frictional Materials*, 5:627–652, 2000.

[16] D. Kolymbas. An outline of hypoplasticity. *Arch. Appl. Mech.*, 61:143–151, 1991.

[17] C. Goldenberg and I. Goldhirsch. Force chains, microelasticity, and macroelasticity. *Phys. Rev. Lett.*, 89(8):084302, 2002.

[18] M. E. Cates, J. P. Wittmer, J.-P. Bouchaud, and P. Claudin. Jamming, force chains, and fragile matter. *Phys. Rev. Lett.*, 81(9):1841–1844, 1998.

[19] H. J. Tillemans and H. J. Herrmann. Simulating deformations of granular solids under shear. *Physica A*, 217:261–288, 1995.

[20] F. Kun and H. J. Herrmann. A study of fragmentation processes using a discrete element method. *Comput. Methods Appl. Mech. Engrg.*, 138:3–18, 1996.

[21] F. Alonso-Marroquin and H.J. Herrmann. Calculation of the incremental stress-strain relation of a polygonal packing. *Phys. Rev. E*, 66:021301, 2002. cond-mat/0203476.

[22] C. Moukarzel and H. J. Herrmann. A vectorizable random lattice. *Journal of Statistical Physics*, 68(5/6):911–923, 1992.

[23] A. Pena, A. Lizcano, F. Alonso-Marroquin, and H. J. Herrmann. Numerical simulations of biaxial test using non-spherical particles. in preparation, 2004.

[24] F. Darve, E. Flavigny, and M. Meghachou. Yield surfaces and principle of superposition: revisit through incrementally non-linear constitutive relations. *International Journal of Plasticity*, 11(8):927, 1995.

[25] F. Alonso-Marroquin, S. Luding, and H.J. Herrmann. Micro-mechanical analysis of the incremental response of a polygonal packing. in preparation, 2004.

[26] F. Alonso-Marroquin, H. J. Herrmann, and S. Luding. Analysis of the elasto-plastic response of a polygonal packing. In *Proceedings ASME (2002)-32498*, New Orleans, 2002. cond-mat/0207698.

[27] F. Alonso-Marroquin and H. J. Herrmann. Ratcheting of granular materials. *Phys. Rev. Lett.*, 92(5):054301, 2004.

[28] R. W. Sharp and J. R. Booker. Shakedown of pavements under moving surface loads. *Journal of Transportation Engineering*, 110:1–14, 1984.

[29] Carl E. Pearson. *Theoretical Elasticity*. Harvard univerity Press, 1952. Harvard Monographs in Applied Science 6.

[30] N. G. W. Cook and K. Hodgson. Some detailed stress strain curves for rocks. *J. Geophys. Research*, 70:2883–2888, 1965.

[31] K. H. Roscoe. The influence of the strains in soil mechanics. *Geotechnique*, 20(2):129–170, 1970.

[32] K. H. Roscoe and J. B. Burland. On the generalized stress-strain behavior of 'wet' clay. In *Engineering Plasticity*, pages 535–609, Cambridge, 1968. Cambridge University Press.

[33] R. Scott. Constitutive relations for soils: Present and future. In *Constitutive Equations for Granular Non-cohesive Soils*, pages 723–726. Balkema, 1988.

[34] D. Kolymbas. *The misery of constitutive modelling*, pages 11–24. Springer, 2000.

[35] M. D. Bolton. The role of micro-mechanics in soil mechanics. In *International Workshop on Soil Crushability*, Japan, 2002. Yamaguchi University.

[36] I. Herle and G. Gudehus. Determination of parameters of a hypoplastic constitutive model from properties of grain assemblies. *Mechanics of Cohesive-Frictional Materials*, 4:461–486, 1999.

[37] P. A. Cundall and O. D. L. Strack. A discrete numerical model for granular assemblages. *Géotechnique*, 29:47–65, 1979.

[38] M. Jean and J. J. Moreau. Unilaterality and dry friction in the dynamics of rigid body collections. In *Proceedings of Contact Mechanics International Symposium*, pages 31–48, Lausanne, Switzerland, 1992. Presses Polytechniques et Universitaires Romandes.

[39] M. P. Allen and D. J. Tildesley. *Computer Simulation of Liquids*. Oxford University Press, Oxford, 1987.

[40] J. P. Bardet. Numerical simulations of the incremental responses of idealized granular materials. *Int. J. Plasticity*, 10:879–908, 1994.

[41] A. V. Potapov and C. S. Campbell. A fast model for the simulation of non-round particles. *Granular Matter*, 1(1):9–14, 1998.

[42] L. Rothenburg and A. P. S. Selvadurai. A micromechanical definition of the Cauchy stress tensor for particulate media. In *Mechanics of Structured Media*, pages 469–486. Elsevier, Amsterdam, 1981.

[43] K. Bagi. Microstructural stress tensor of granular assemblies with volume forces. *J. Appl. Mech.*, 66:934–936, 1999.

[44] S. Luding and H. J. Herrmann. Micro-macro transition for cohesive granular media. In *Bericht Nr. II-7*, pages 121–133. Inst. für Mechanik, Universität Stuttgart, 2001.

[45] N. P. Kruyt and L. Rothenburg. Micromechanical definition of strain tensor for granular materials. *ASME Journal of Applied Mechanics*, 118:706–711, 1996.

[46] K. Bagi. Stress and strain in granular assemblies. *Mech. of Materials*, 22:165–177, 1996.

[47] F. Calvetti, C. Tamagnini, and G. Viggiani. On the incremental behaviour of granular soils. In *Numerical Models in Geomechanics*, pages 3–9, Lisse, 2002. Swets & Zeitlinger.

[48] Y. Kishino. On the incremental nonlinearity observed in a numerical model for granular media. *Italian Geotechnical Journal*, 3:3–12, 203.

[49] F. Calvetti, G. Viggiani, and C. Tamagnini. Micromechanical inspection of constitutive modelling. In *Constitutive modelling and analysis of boundary value problems in Geotechnical Engineering*, pages 187–216., Benevento, 2003. Hevelius Edizioni.

[50] F. Kun and H. J. Herrmann. Transition from damage to fragmentation in collision of solids. *Phys. Rev. E*, 59(3):2623–2632, 1999.

[51] A. Okabe, B. Boots, and K. Sugihara. *Spatial Tessellations. Concepts and Applications of Voronoi Diagrams*. John Wiley & Sons, Chichester, 1992. Wiley Series in probability and Mathematical Statistics.

[52] F. Kun, G. A. D'Addetta, E. Ramm, and H. J. Herrmann. Two-dimensional dynamic simulation of fracture and fragmention of solids. *Comp. Ass. Mech. Engng.*, 6:385–402, 1999.

[53] T. Marcher and P. A. Vermeer. Macromodelling of softening in non-cohesive soils. In *Continuous and Discontinuous Modelling of Cohesive Frictional Materials*, pages 89–110, Berlin, 2001. Springer.

[54] J. A. Astrøm, H.J.Herrmann, and J. Timonen. Granular packings and fault zones. *Phys. Rev. Lett.*, 84:4638–4641, 2000.

[55] E. Buckingham. On physically similar systems: Illustrations of the use of dimensional equations. *Phys. Rev.*, 4:345–376, 1914.

[56] I. Vardoulakis and J. Sulem. *Bifurcation analysis in geomechanics*. hapman & Hall, London, 1995.

[57] P. A. Cundall. Numerical experiments on localization in frictional materials. *Ingenieur-Archiv*, 59:148–159, 1989.

[58] M. Lätzel, S. Luding, and H. J. Herrmann. Macroscopic material properties from quasi-static, microscopic simulations of a two-dimensional shear-cell. *Granular Matter*, 2(3):123–135, 2000. cond-mat/0003180.

[59] M. Oda and H. Kazama. Microstructure of shear bands and its relation to the mechanism of dilatancy and failure of dense granular soils. *Geotechnique*, 48(4):465–481, 1998.

[60] A. Casagrande. Characteristics of cohesionless soils affecting the stability of slopes and earth fills. *J. Boston Soc. Civil Eng.*, pages 257–276, 1936.

[61] L.D. Landau and E. M. Lifshitz. *Theory of Elasticity*. Pergamon Press, Moscou, 1986. Volume 7 of Course of Theoretical Physics.

[62] M. Oda and K. Iwashita. Study on couple stress and shear band development in granular media based on numerical simulation analyses. *Int. J. of Enginering Science*, 38:1713–1740, 2000.

[63] G. Gudehus. A comparison of some constitutive laws for soils under radially symmetric loading and unloading. *Can. Geotech. J.*, 20:502–516, 1979.

[64] D.C. Drucker and W. Prager. Soil mechanics and plastic analysis of limit design. *Q. Appl. Math.*, 10(2):157–165, 1952.

[65] R. Nova and D. Wood. A constitutive model for sand in triaxial compression. *Int. J. Num. Anal. Meth. Geomech.*, 3:277–299, 1979.

[66] P. A. Vermeer. A five-constant model unifying well-established concepts. In *Constitutive Relations of soils*, pages 175–197, Rotterdam, 1984. Balkema.

[67] H. B. Poorooshasb, I. Holubec, and A. N. Sherbourne. Yielding and flow of sand in triaxial compression. *Can. Geotech. J.*, 4(4):277–398, 1967.

[68] E. M. Wood. *Soil Mechanics-transient and cyclic loads*. Chichester, 1982.

[69] Y. F. Dafalias and E. P. Popov. A model of non-linearly hardening material for complex loading. *Acta Mechanica*, 21:173–192, 1975.

[70] R. Chambon, J. Desrues, W. Hammad, and R. Charlier. CLoE, a new rate type constitutive model for geomaterials. Theoretical basis and implementation. *Int. J. Anal. Meth. Geomech.*, 18:253–278, 1994.

[71] W. Wu, E. Bauer, and D. Kolymbas. Hypoplastic constitutive model with critical state for granular materials. *Mech. Matter.*, 23:45–69, 1996.

[72] G. Gudehus. Attractors, percolation thresholds and phase limits of granular soils. In *Powders & Grains 97*, pages 169–183, 1997.

[73] D. Kolymbas. *Modern Approaches to Plasticity.* Elsevier, 1993.

[74] A. Drescher and G. de Josselin de Jong. Photoelastic verification of a mechanical model for the flow of a granular material. *J. Mech. Phys. Solids*, 20:337–351, 1972.

[75] H. M. Jaeger and S. R. Nagel. Granular solids, liquids and gases. *Rev. Mod. Phys.*, 68:1259, 1996.

[76] P. Dubujet and F. Dedecker. Micro-mechanical analysis and modelling of granular materials loaded at constant volume. *Granular Matter*, 1(3):129–136, 1998.

[77] Y. F. Dafalias. Bounding surface plasticity. I: Mathematical foundation and hypoplassticity. *J. of Engng. Mech*, 112(9):966–987, 1986.

[78] Tatsouka F and K. Ishihara. Yielding of sand in triaxial compression. *Soils and Fundations*, 14(2):63–76, 1974.

[79] R. Hill. A general theory of uniqueness and stability in elastic-plastic solids. *Journal of Geotechnical Engineering*, 6:239–249, 1958.

[80] M. A. Stroud. *The behavior of sand at low stress levels in the simple shear appatus.* PhD thesis, University of Cambridge, 1971.

[81] H.-B. Mühlhaus and I. Vardoulakis. The thickness of shear bands in granular materials. *Géotechnique*, (37):271–283, 1987.

[82] S. Torquato. Exact expression for the effective elastic tensor of disordered composites. *Phys. Rev. Lett.*, 79(4):681–684, 1997.

[83] G. Festag. Experimental investigation on sand under cyclic loading. In *Constitutive and Centrifuge Modelling: two Extremes*, pages 269–277, Monte Verita, 2003.

[84] G. Festag. Experimenelle und numerische Untersuchungen zum Verhalten von granularen Materialien unter zyklischer Beanspruchung. Dissertation TU Darmstadt, 2003.

[85] G. Gudehus. Ratcheting und DIN 1054. *10. Darmstädter Geotechnik-Kolloquium*, 64:159–162, 2003.

[86] F. Lekarp, A. Dawson, and U. Isacsson. Permanent strain response of unbound aggregates. *J. Transp. Engrg.*, 126(1):76–82, 2000.

[87] S. Khedr. Deformation characteristics of granular base course in flexible pavement. *Transportation Research Record*, 1043:131–138, 1985.

[88] R. Lentz. *Permanent deformation for cohesionless subgrade materials under cyclic loading*. PhD thesis, Michigan State University, East Lansing, 1979.

[89] S. Werkmeister, A. R. Dawson, and F. Wellner. Permanent deformation behavior of granular materials and the shakedown theory. *Journal of Transportation Research Board*, 1757:75–81, 2001.

[90] F. Lekarp and A. Dawson. Modelling permanent deformation behaviour of unbound granular materials. *Construction and Building Materials*, 12(1):9–18, 1998.

[91] R. Barksdale. Laboratory evaluation of rutting in base course materials. In *Third International Conference on Structural Design of Asphalt Pavements*, pages 161–174, London, 1972.

[92] R.P.Feynman, R. B. Leighton, and M. Sands. *The Feynman Lectures on Physics*, chapter 46. Addison - Wesley.

[93] J. Howard. Molecular motors: Structural adaptation to cellular functions. *Nature*, 389:561, 1997.

[94] P. Reimann. Brownian motors: Noisy transport far from equilibrium. *Phys. Rep.*, 361:57, 2002.

[95] Farkas, P. Tegzes, A. Vukics, and T. Vicsek. Transitions in the horizontal transport of vertically vibrated granular layers. *Phys. Rev. E*, 60:7022, 1999.

[96] Z. Farkas, F. Szalai, D. E. Wolf, and T. Vicsek. Segregation of granular binary mixtures by a ratchet mechanism. *Phys. Rev. E*, 65:022301, 2002.

[97] J.F. Wambaugh, C. Reichhardt, and C.J. Olson. Ratchet-induced segregation and transport of nonspherical grains. *Phys. Rev. E*, 65(022301), 2002.

[98] G. Sweere. *Unbound granular bases for roads*. PhD thesis, University of Delf, 1990.

[99] D. Kolymbas, I. Herle, and P. A. Wollferdorff. A hypoplastic constitutive equation with back stress. *Int. J. Anal. Meth. Geomech.*, 19:415–446, 1995.

[100] A. Niemunis and I. Herle. Hypoplastic model for cohesionless soils with elastic strain range. *Int. J. Mech. Cohesive-Frictional Mater.*, 2:279–299, 1996.

[101] I. Ishibashi and H. Kiku. Effect of initial anisotropy on liquefaction potential by discrete element model. In *First International Conference on Earthquake Geotechnical Engineering (IS-TOKYO '95)*, volume II, pages 863–868, Rotterdam, 1995. Balkema.

[102] M. Nicolas, P. Duru, and O. Pouliquen. Compaction of a granular material under cyclic shear. *Eur. Phys. J. E*, 3(4):309, 2000.

[103] A. J. Liu and S.R. Nagel. Jamming is not just cool any more. *Nature*, 396:21, 1998.

[104] B. Doliwa and A. Heuer. Cage effect, local anisotropies, and dynamic heterogeneities at the glass transition: A computer study of hard spheres. *Phys. Rev. Lett.*, 80:4915–4918, 1998.

[105] L .E. Silbert, D. Ertas, G. S. Grest, T.C. Halsey, and D. Levine . Analogies between granular jamming and the liquid-glass transition. *Physical Review E*, 65:051307, 2002.

[106] M. D. Rintoul and S. Torquato. Metastability and crystallization in hard-sphere systems. *Phys. Rev. Lett.*, 77(20):4198–4201, 1996.

[107] M. Nicodemi, A. Coniglio, and H. J. Herrmann. A model for the compaction of granular media. *Physica A*, 225:1–6, 1995.

[108] E. Caglioti, V. Loreto, H. J. Herrmann, and M. Nicodemi. A "Tetris-like" model for the compaction of dry granular media. *Phys. Rev. Lett.*, 79(8):1575–1578, 1997.

[109] F. Radjai and S. Roux. Turbulent-like fluctuations in quasistatic flow of granular media. *Phys. Rev. Lett.*, 89(6):064302, 2002.

[110] H.-B. Mühlhaus and P. Hornby. On the reality of antisymmetric stresses in fast granular flows. In *IUTAM Symposium on Mechanics of Granular and Porous Materials*, pages 299–311. Kluwer Academic Publishers, 1997.

[111] I. Vardoulakis. Shear-banding and liquefaction in granular materials on the basis of a Cosserat continuum theory. *Ingenieur-Archiv*, 59:106–113, 1989.

[112] F. Alonso-Marroquin, Garcia-Rojo, and H. J. Herrmann. Micromechanical investigation of granular ratcheting. In *Proceeding of International Conference on Cyclic Behaviour of Soils and Liquefaction Phenomena*, Bochum-Germany, 2004. cond-mat/0207698.

[113] F. Alonso-Marroquin, S. McNamara, and H.J. Herrmann. Micromechanische Untersuchung des granulares Ratchetings. *DFG Antrag*, 2003.

[114] G. R. McDowell, M. D. Bolton, and D. Robertson. The fractal crushing of granular materials. *J. Mech. Phys. Solids*, 44(12):2079–2102, 1996.

[115] D. Coppersmith. Model for force fluctuations in bead packs. *Phys. Rev. E*, 53:4673–4685, 1996.

[116] S. Roux and F. Radjai. On the state variables of the granular materials. In *Mechanics of a New Millenium*, pages 181–196, Kluwer, Dordrecht, 2001.

Acknowledgment

At the end of this thesis I would like to express deepest gratitude to all those who made this work possible.

First, I want to thank Prof. Dr. Hans J. Herrmann who gave me the opportunity to work on my thesis at the ICA1, who offered me excellent guidance and a continuous encouragement and support during this study.

Next, I am indebted to Prof. Dr-Ing. P. Vermeer who accepted to act as a referee of this thesis.

I am grateful to F. Darve, who inspired many theoretical questions in soil mechanics, and for his fruitful collaboration.

I express my gratitude to P. Cundall, C. Detournay and E. Detournay for their hospitality during my visit to Itasca and the University of Minnesota.

I thank G. Gudehus and A. Schuenemann, for the enlightening discussions on the micromechanical aspects of soil plasticity. Without their valuable remarks and stimulation, the detection of the *granular ratcheting* would have been impossible.

I want to express my thanks to the members of the DIGA project, especially I. Vardoulakis, M. Pastor, R. Nova, G. Pijaudier-Cabot and F. Darve for providing me an excellent introduction to the area of Geomaterials in the CISM training course in Udine.

I thank E. Ramm, D. Ehlers, P. Vermeer, S. Luding, S. Wenz, G.A. D'Addetta U. Vogler, M. Latzel and T. Marcher for helpful discussions inside of the research group *Modelling of Cohesive Frictional Materials*.

I would like to thank those people who contributed many valuable ideas to this work, especially to J. A. Astrøm, K. Bagi, T. Benz, F. Calvetti, P. Cundall, R. Chambon, Y. F. Dafalias, F. Froiio, G. Festag, J. Gallas, Y. Kishino, F. Kun, D. Kolymbas, S. Luding, A. Lizcano, S. McNamara, A.

Niemunis, A. Peña, D. Potyondy, T. Unger, G. Viggiani, D. Wood and other people whose names escape my memory at this moment.

I am also grateful to the members of the ICA1, Stuttgart, who indirectly supported this work by an excellent working atmosphere. A very special thanks to R. C. Hidalgo, R. Garcia-Rojo, J.D. Muñoz for many useful discussions and their constructive comments; to Frank Huber, for his constant support as system administrator, to Marta González, Eric Riveiro Partelli, Alejandro Mora, Frank Fonseca, Veit Schwämmle and the philosopher Adriano de Oliviera Sousa for the nice coffee sessions. My special thanks also go to Vasanthi, Reza Mahmoodi, Frank Raischel and Sean McNamara for tireless in writting corrections to this work; and Marlies Parsons who supported me at the ICA1 concerning all administrative and accommodation affairs.

Finally, I would like to express my gratitude for the constant support of all the people I have been associated with during the last years, especially the *Realkücheverein* (OLA) for the wonderful meals, *tertulias* and desserts, and the unconditional friendships of its members.

I acknowledge the support of this work by the Deutsche Forschungsgemeinschaft (DFG) within the Research group *Modellierung kohäsiver Reibungsmaterialen* and the European Union project *Degradation and Instabilities of Geomaterials with Application to Hazard Mitigation* (DIGA) in the framework of the Human Potential Program, Research Training Networks (HPRN-CT-2002-00220).

Un especial agradecimiento a mis padres y a mis hermanos, quienes siempre estuvieron a mi lado pese a la enorme distancia que nos separa.

Dedicado a Nicola, a Amandita y a nuestro pequeño(a) griego(a), como un tributo al profundo amor que nos une.

Fernando Alonso Marroquín
Stuttgart, April 21, 2004